AURORA BOREALIS

The Amazing Northern Lights

Volume 6, Number 2

COVER — *The winter night gleams with an auroral display over a log house in the Goldstream Valley, near Fairbanks. (G. Lamprecht, Geophysical Institute, University of Alaska)*

RIGHT — *Fisheye-lens ("all-sky") photograph of the aurora taken from the University of Alaska campus, Fairbanks. (Geophysical Institute, University of Alaska)*

ALASKA GEOGRAPHIC®

Copyright © 1979 The Alaska Geographic Society. All rights reserved.

An impressive display of the aurora borealis, sketched by M. Bravais during his 1837-38 expedition in northern Norway. (Courtesy of S.-I. Akasofu)

The Alaska Geographic Society

To teach many more to better know and use our natural resources

About This Issue: The editors of The Alaska Geographic Society are fortunate to have one of the world's authorities on auroras as the author of this issue. S.-I. Akasofu received his Ph.D. at the University of Alaska and is now professor of geophysics at the Geophysical Institute, University of Alaska, Fairbanks. Dr. Akasofu has many publications to his credit, including two books on magnetospheric substorms, *Polar and Magnetospheric Substorms* and *Physics of Magnetospheric Substorms.* In addition, he is the coauthor of *Solar-Terrestrial Physics.* Dr. Akasofu is the recipient of the Chapman Medal from the Royal Astronomical Society, England, and the Japan Academy Award for his work on magnetospheric physics.

We are also grateful to the Geophysical Institute, University of Alaska, Fairbanks, for their cooperation and assistance with this project. The institute was established by the U.S. Congress in the late 1940's as a center for geophysical research in the Arctic and its staff is involved in research on auroras, magnetic storms, radio communications, climate, air pollution, glaciers, earthquakes and many other phenomena that affect the arctic environment.

The editors also wish to acknowledge the assistance of scientists and others who have shared their photographs, artwork and other materials relating to the auroras. To the National Science Foundation, which has supported Dr. Akasofu's work; to the staff of the Elmer Rasmuson Library, University of Alaska, Fairbanks, which assisted with research; to Dorothy Jean Ray, author of many Eskimo life books, for her contribution to old legends; to the family of the late E.L. Keithahn, one time curator of the Alaska Territorial Museum, for a piece he once wrote for the old *ALASKA SPORTSMAN®* on Native stories about the aurora; and to Nancy Walters, who assisted the author in research for this project, we extend a special thanks.

Editors: Robert A. Henning, Marty Loken, Barbara Olds
Associate Editor: Penny Rennick
Designer: Roselyn Pape
Illustrator: Jon. Hersh

ALASKA GEOGRAPHIC®, ISSN 0361-1353, is published quarterly by The Alaska Geographic Society, Anchorage, Alaska 99509. Second-class postage paid in Edmonds, Washington 98020. Printed in U.S.A.

THE ALASKA GEOGRAPHIC SOCIETY is a nonprofit organization exploring new frontiers of knowledge across the lands of the polar rim, learning how other men and other countries live in their Norths, putting the geography book back in the classroom, exploring new methods of teaching and learning—sharing in the excitement of discovery in man's wonderful new world north of 51°16'.

MEMBERS OF THE SOCIETY RECEIVE *Alaska Geographic®*, a quality magazine in color which devotes each quarterly issue to monographic in-depth coverage of a northern geographic region or resource-oriented subject.

MEMBERSHIP DUES in The Alaska Geographic Society are $20 per year. (Eighty percent of each year's dues is for a one-year subscription to *Alaska Geographic®*.) Order from The Alaska Geographic Society, Box 4-EEE, Anchorage, Alaska 99509; (907) 274-0521.

MATERIAL SOUGHT: The editors of *Alaska Geographic®* seek a wide variety of informative material on the lands north of 51°16' on geographic subjects—anything to do with resources and their uses (with heavy emphasis on quality color photography)—from Alaska, Northern Canada, Siberia, Japan—all geographic areas that have a relationship to Alaska in a physical or economic sense. (In early 1979 editors were seeking photographs and other materials on the following subjects: Stikine River drainage; shellfish and shellfisheries of Alaska; Aleutian Islands; Yukon River and its tributaries; Wrangell and Saint Elias Mountains; and Alaska's Great Interior.) We do not want material done in excessive scientific terminology. A query to the editors is suggested. Payments are made for all material upon publication.

CHANGE OF ADDRESS: The post office does not automatically forward *Alaska Geographic®* when you move. To insure continuous service, notify us six weeks before moving. Send us your new address and zip code (and moving date), your old address and zip code, and if possible send a mailing label from a copy of *Alaska Geographic®*. Send this information to *Alaska Geographic®* Mailing Offices, 130 Second Avenue South, Edmonds, Washington 98020.

MAILING LISTS: We have begun making our members' names and addresses available to carefully screened publications and companies whose products and activities might be of interest to you. If you would prefer not to receive such mailings, please so advise us, and include your mailing label (or your name and address if label is not available).

Registered Trademark: *Alaska Geographic*. Library of Congress catalog card number 72-92087.
ISSN 0361-1353; key title *Alaska Geographic*.
ISBN 0-88240-124-6

The entire sky of Interior Alaska is often bathed in auroras such as this one. (A. Lee Snyder, Jr., Geophysical Institute, University of Alaska)

Back in the early thirties at the old Alaska Agricultural College and School of Mines, near Fairbanks, the late Professor Veryl R. Fuller, one of the early researchers in auroral phenomena, used to supervise auroral watches as our Alaskan contribution to a government-funded International Polar Year. Watchers around the global polar rim arranged to take photos simultaneously with a specially developed Eastman photographic film of extreme sensitivity. College students were paid to keep nightlong vigils under the constantly expanding or shrinking aurora, squeezing the bulb on the big view camera on the appointed hours. It was bitter cold work, sometimes fifty below, but always fascinating. Different forms, wavering, many colors diffusing and changing, sometimes far away, sometimes filling the heavens around and above, plunging great dropping spears and sheets of color earthward toward your very head as though a great hand were dropping color like burning oil. At such times you could feel yourself involuntarily cringe. An unforgettable experience. Charged ions, sure, and all the other scientific findings—but is all this beauty of the night so simply explained? Holy? Supernatural? Scientifically explainable, they say. They still make us feel spooky. There are those—and my wife is among them—who claim on a stack of Bibles they can hear them. Dr. Fuller, wherever you are, I hope you get a chance to look at this issue.

Robert A. Henning

President
The Alaska Geographic Society

Introduction

But, where, O Nature, is thy law?
From the midnight lands comes up the dawn!
Is it not the sun setting his throne?
Is it not the icy seas that are flashing fire?
Lo, a cold flame has covered us!
Lo, in the night-time day has come upon the earth.

What makes a clear ray tremble in the night?
What strikes a slender flame into the firmament?
Like lightning without storm clouds,
Climbs to the heights from earth?
How can it be that frozen steam
Should midst winter bring forth fire?

M. V. Lomonosov, *Aurora and Airglow*, ed. by
B. M. McCormac, translated by K. Chapman,
New York: Reinhold, 1967, p. 20

The aurora borealis—the northern lights—is one of the most spectacular natural phenomena on earth. Its beauty and splendor are often beyond description. Thus, Charles F. Hall, an author and polar explorer of the last century, simply exclaimed with a sigh, "Who but God can conceive such infinite scenes of glory? Who but God could execute them, painting the heavens in such gorgeous display?"

William H. Hooper, another polar explorer, reported, "Language is vain in the attempt to describe its ever varying and gorgeous phases; no pen nor pencil can portray its fickle hues, its radiance, and its grandeur." Even so, many of those characters passing through the history of the arctic and antarctic regions, among them adventurers, scholars, miners, settlers and gamblers, made an attempt to communicate in words to their less fortunate fellows the wonder they witnessed. Their writings are frequently full of interest to us today.

Although no description of the aurora can do it full justice, its magnificence is at least suggested in a poem by the 18th century Russian scientist Mikhail Vasil'evich Lomonosov.

Few today have not become acquainted to some extent with tales of the lights in the cold northern sky, perhaps from published accounts of polar explorers and adventurers, from some traveler friend who has seen them firsthand or from a striking picture seen in a magazine. This book is an attempt to present the whole story of the aurora.

The aurora borealis has intrigued mankind since ancient

A woodcut by Fridtjof Nansen shows his ship, the Fram, *in the arctic pack ice, framed in a great curl of the aurora borealis. (From Nansen's* Nord I Takeheimen, *1911, courtesy of Auroral Observatory, University of Tromsø, Norway)*

CLOCKWISE FROM RIGHT — *Many Eskimos believed the aurora was alive and that if a person whistled at it, it would come closer out of curiosity. An active aurora of this type is often referred to as an active rayed band. (S.-I. Akasofu) / Some Eskimos imagined the aurora might be torches lit by spirits to guide travelers on their journey to heaven. This aurora is an example of a slightly active band, developing a curl structure. (T. Ohtake, Geophysical Institute, University of Alaska) / The silence of dazzling waves of auroral light — which seem worthy of accompanying thunder — can evoke strange feelings in the viewer of the night sky. (Geophysical Institute, University of Alaska)*

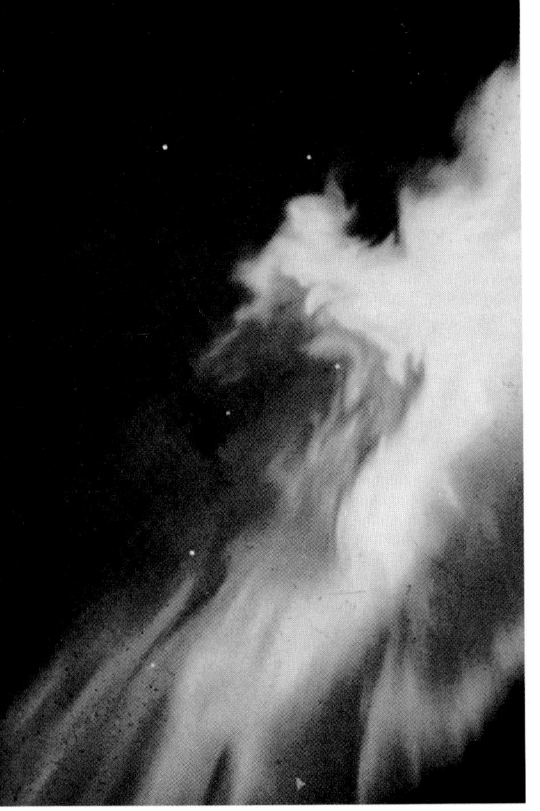

times. Early descriptions of the aurora are contained in the Old Testament, in the mythology of the Lapps, the Eskimos and the American Indians as well as in Medieval European literature. After sampling some of these accounts, we shall proceed to the narratives of polar explorers and tales of settlers in the northern lands.

In its technical aspects, the aurora has fascinated many famous philosophers and scientists, among them, Aristotle, Descartes, Goethe, Henry Cavendish, John Dalton, Edmund Halley, Alfred L. Wegener and Benjamin Franklin. It has provided one of the most challenging problems encountered in modern science. In these pages, you will be introduced to the history of auroral science and the development of our understanding of the phenomenon, and we'll explain, in layman's language, what we now know about how the aurora works. Finally, the reasons for auroral study today will be outlined, with particular emphasis on its great relevance to our technological future.

A medieval artist's view of the aurora borealis over the Bavarian city of Bamberg, about 1560. Active auroras, appearing often as swift streams of light, were associated with battles fought in the sky. (Department of Prints and Drawings, Zentralbibliothek, Zurich)

Auroral Legends

One can find scientific descriptions of the aurora as far back as the days of the Roman Empire, in Seneca's *Questiones Naturales*, and later in Aristotle's *Meteorologica*. Seneca, a Roman philosopher at the beginning of the Christian era, noted:

> "There are chasmata [fissures], when a certain portion of the sky opens, and gaping displays the flame as in a porch. The colours also of all these are many. Certain are of the brightest red, some of a flitting and light flame-colour, some of a white light, others shining, some steadily and yellow without eruptions or rays."

There are a number of biblical references to phenomena which can be attributed to auroral activity. In the Old Testament one example is in 2 Maccabees, chapter 5, verses 1-4, written about 176 B.C.:

> About this time Antiochus sent his second expedition into Egypt. It then happened that all over the city, for nearly forty days, there appeared horsemen charging in midair, clad in garments interwoven with gold — companies fully armed with lances and drawn swords; squadrons of cavalry in battle array, charges and countercharges on this side and that, with brandished shields and bristling spears, flights of arrows and flashes of gold ornaments, together with armor of every sort. Therefore all prayed that this vision might be a good omen.

Virtually every northern culture has its oral legends about the aurora, passed down for generations. The Eskimos, Athabascan Indians, Lapps, Greenlanders and even the Northwest Indian tribes were familiar with this mysterious light in the sky. Their legends took many forms and were most often associated with their notions of life after death. Typical Eskimo stories on the aurora may be found in books by Knud Rasmussen and Ernest W. Hawkes, explorers and anthropologists who set to paper oral traditions. In his book, *The Labrador Eskimo*, Hawkes related one legend:

> The ends of the land and sea are bounded by an immense abyss, over which a narrow and dangerous

pathway leads to the heavenly regions. The sky is a great dome of hard material arched over the earth. There is a hole in it through which the spirits pass to the true heavens. Only the spirits of those who have died a voluntary or violent death, and the raven, have been over this pathway. The spirits who live there light torches to guide the feet of new arrivals. This is the light of the aurora. They can be seen there feasting and playing football with a walrus skull.

The whistling crackling noise which sometimes accompanies the aurora is the voices of these spirits trying to communicate with the people of the earth. They should always be answered in a whispering voice. Youths and small boys dance to the aurora. The heavenly spirits are called selamiut, "sky-dwellers," those who live in the sky.

Ernest W. Hawkes, *The Labrador Eskimo*, Ottawa: Government Printing Bureau, 1916, p. 153

Rasmussen's story follows the same general theme, even though it is taken from another Eskimo culture.

The dead suffer no hardship, wherever they may go, but most prefer nevertheless to dwell in the Land of Day, where the pleasures appear to be without limit. Here, they are constantly playing ball, the Eskimos' favourite game, laughing and singing, and the ball they play with is the skull of a walrus. The object is to kick the skull in such a manner that it always falls with the tusks downwards, and thus sticks fast in the ground. It is this ball game of the departed souls that appears as the aurora borealis, and is heard as a whistling, rustling, crackling sound. The noise is made by the souls as they run across the frost-hardened snow of the heavens. If one happens to be out alone at night when the aurora borealis is visible, and hears this whistling sound, one has only to whistle in return and the light will come nearer, out of curiosity.

Knud Rasmussen: *Intellectual Culture of the Iglulik Eskimo*. Fifth Thule Expedition. Copenhagen: Glydem Dalski Boghand, 1932, p. 95

American Indians, too, have very interesting legends about the aurora.

Chief M'Sartto, Morning Star, had an only son, different from all others in the tribe. The son would not play with other children but would take his bow and arrow and be away for days. Curious as to what mischief his boy could be up to, the Chief one day followed him. His journey progressed and all at once a queer feeling filled the old Chief, as if all knowledge was floating away from him. His eyes suddenly closed and when they were opened he found himself in an extremely light country with no sun, moon, or stars. There were many people about, and they spoke a strange language which he did not understand. The people were engaged in a wonderful game of ball which seemed to turn the light to many colors. The players all had lights on their heads and wore very curious belts called Menquan, or Rainbow belts. After many days of searching for his son, the old Chief met a man who

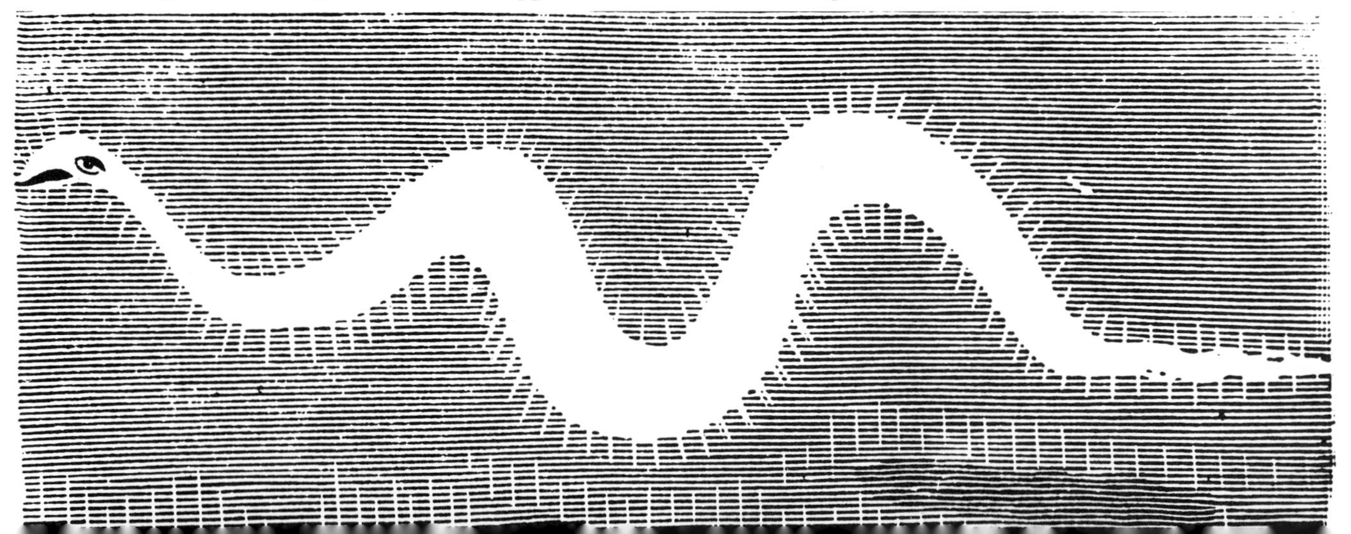

Some people once thought of the aurora as a radiating snake dancing in the sky. (From The Cry in the Midnight; See the Bridegroom Coming *by E.M. Handie, 1883, courtesy of S.-I. Akasofu)*

Another typical medieval view of the aurora — troops of horsemen running in the sky. (N. Pushkov and S. I. Isaev)

spoke his language. The man had also traveled by chance to this strange new country and knew of the Chief's son. When he was brought to his son, the Chief saw him playing ball with the others, and strangely enough the boy's light was brighter than any there. When the game was ended, the old Chief was introduced to the people and honored by the Chief of the Northern Lights; two great birds were ordered to be brought forth, K'che Sippe by name. On these birds the two dwellers of the Lower World returned home from the Wa-ba-ban, the land of the Northern Lights, following the Spirits Path, Ket-a-gus-wowt or Milky Way. Again all knowledge was erased from Chief M'Sartto. When they arrived home, the Chief's wife paid no notice that they were gone because she was afraid that they would never return and was very relieved. So it is that those very few who travel to the Land of the Northern Lights do not remember their remarkable journey. But the Chief's son remembers and travels there often.

Katharine B. Judson, *Myths and Legends of the Pacific Northwest*. Chicago: A. C. McClurg and Co., 1910, pp. 143-5

In the 13th century Norwegian chronicle, the *King's Mirror* (a Viking story), we find the king instructing his son about the aurora.

No man is sure what those lights can be which the Greenlanders call the Northern Lights, even those who have spent a long time in Greenland. Of course, thoughtful men will make conjectures as to what they might consist of. Their nature is peculiar in that the darker the night is, the brighter they seem and they always appear at night, never by day. In appearance they resemble a vast flame of fire viewed from a great distance. It also looks as if sharp points were shot from this flame up into the sky of uneven height and in constant motion, now one, now another darting highest; and the light appears to blaze like a living flame. When these rays are at their highest, people out of doors can easily find their way about and can even go hunting. Sometimes, however, the light appears to grow dim, as if a black smoke or a dark fog were blown up among the rays; and then it looks very much as if the light were overcome by this smoke and about to be quenched. It happens at times that people think they see large sparks shooting out of the light from glowing

iron which has just been taken from the forge. As night declines, the light begins to fade and when daylight appears, it seems to vanish entirely.

Men who have thought about the origin of these lights have guessed at three sources, one of which ought to be the true one. Some hold that fire circles about the ocean and all the bodies of water that stream about on the outer sides of the globe; since Greenland lies on the outermost edge of the earth to the north, they think it possible that these lights shine forth from the fires that encircle the outer ocean. Others say that during the hours of night, when the sun's course is beneath the earth, an occasional gleam of its light may shoot up into the sky; for they insist that Greenland lies so far out on the earth's edge that the curved surface which shuts out the sunlight must be less prominent there. Still others believe that the frost and the glaciers have become so powerful there that they are able to radiate forth these flames.

The King's Mirror (Speculum Regale - *Konungs Skuggsja*), translated from the old Norwegian by Laurence Marcellus Larson, New York: The American-Scandinavian Foundation, 1917, pp. 146-51

On many occasions in historical times the aurora has reached as far south as the middle latitudes, and it has struck fear into the populations of Italy and France. The mid-latitude aurora is a rich dark red color, and the people of Europe associated it with blood and battle; to them it was an ill omen of disasters to come. Accounts of such red auroras can be found in some of the earliest of Eastern and Western writings. A typical Chinese description of the red aurora is of a "red cloud spreading all over the sky, and among the red cloud there were ten-odd bands of white vapour like glossed silk penetrating it."

Seneca remarked, "Amongst these we may notice, what

LEFT — *The red aurora is one of the most spectacular displays of nature. This one was photographed in Fairbanks on February 11, 1958, and was also seen across the Lower 48 and Canada. Since the days of the Roman Empire, this type of aurora has often been mistaken for a huge fire. Even in 1958 there were reports of fire engines rushing out to extinguish the red aurora. (V.P. Hessler, Geophysical Institute, University of Alaska)*

The aurora figured prominently in the legends of the Indian tribes of the North. (Geophysical Institute, University of Alaska)

we frequently read of in history, the sky is seen to burn, the glow of which is occasionally so high that it may be seen amongst the stars themselves, sometimes so near the earth that it assumes the form of a distant fire. Under Tiberius Caesar the cohorts ran together in aid of the colony of Ostia as if it were in flames, when the glowing of the sky lasted through a great part of the night, shining dimly like a vast and smoking fire."

In his monograph, *The Aurora Borealis*, Alfred Angot, honorary meteorologist at the central meteorological office of France, described the terror caused by red auroras in medieval times.

> . . . astrology had so troubled the minds of men that the aurora borealis had become a source of terror: bloody lances, heads separated from the trunk, armies in conflict, were clearly distinguished. At the sight of them people fainted (according to Cornelius Gemma), others went mad. Pilgrimages were organised to avert the wrath of Heaven, manifested by these terrible signs. Thus, according to the Journal of Henri III, in the month of September 1583 eight or nine hundred persons of all ages and both sexes, with their lords, came to Paris in procession, dressed like penitents or pilgrims, from the villages of Deux-Gemeaux and Ussy-en-Brie, near La Ferte-Gaucher, "to say their prayers and make their offerings in the great church at Paris; and they said that they were moved to this penitential journey because of signs seen in heaven and fires in the air, even towards the quarter of the Ardennes, whence had come the first such penitents, to the number of ten or twelve thousand, to Our Lady of Rheims and to Liesse." The chronicler adds that this pilgrimage was followed a few days afterwards by five others, and for the same cause.
>
> Alfred Angot, *The Aurora Borealis*, New York: D. Appleton, 1897, p. 6-7

Under the circumstances, it is surprising to find a painting from Medieval days which depicts the aurora as a series of candles in the sky, but it is quite likely that such a romantic view of the aurora was not very common.

A 19th century drawing of a draped aurora over Paris. (From The Aurora Borealis, *by Alfred Angot, 1897, courtesy of S.-I. Akasofu)*

Ein vnerhörtes Wunderzeichen/welches ist gesehen worden auff Kuttenberg/in der Kron Böhem/auch sonst in andern Stätten vnd Flecken herumb/ den 12. Januarij/vier stund in die Nacht/vnd gewehret biß nach 8. Inn der Wolcken des Himels stehen/ alß in disem Jar. 1 5 7 0.

An unusually fanciful interpretation of the aurora of medieval days (Bohemia, 1570). More commonly the appearance of the phenomena instilled terror in the populations of Europe. (Courtesy of The Astronomer Royal of Scotland)

Legends of the Northern Lights

By Dorothy Jean Ray
Reprinted from
The ALASKA SPORTSMAN®
April 1958

How often have you watched them playing across the night sky and wondered, just what are the northern lights?

From the time man first watched the twisting, writhing contortions of mysterious light in the far northern latitudes, he has thought them to be many things, from spirits playing ball with a walrus skull to giant fingers in the sky.

The Eskimos and Indians of North America had many explanations for the aurora borealis, most of them now forgotten except in an occasional ethnographer's or traveler's account. But a few wisps of superstition still cling to the northern lights, and it is not likely that they will be dispelled even by the extensive research being carried on during the International Geophysical Year.

A Disturbing Influence

One of the major stations for the study of the aurora is the Geophysical Institute at the University of Alaska, which is participating with 56 nations and more than 5,000 experts in the general field of geophysics. Although the study of the aurora borealis is only one of the various problems of the earth and its physical forces employing scientists during this time, it is important to the North because of the direct relationship between auroral disturbances and telephone and radio communications. During a violent auroral storm, communications are sometimes virtually cut off.

Many fine descriptions of auroral displays are found in old exploring and sailing stories, but it was not until after the turn of the 20th century, and improvements in the spectroscope and

The auroral curtain often appears in multiple layers. There are at least five curtains in this photograph. (T. Ohtake, Geophysical Institute, University of Alaska)

specialized cameras, that scientific studies were conducted. It was apparent even in the early days of auroral research that the displays were caused by magnetic disturbances from the sun, which produced light when they collided with atoms of the upper air.

In spite of the fact that the cause has been known for a long time, many fantastic explanations of the lights have been nurtured to this day by uninformed persons — that they are explosions occurring when warm and cold air meet, or lost lightning from the earth. Even those who know what the aurora is still ascribe to it impossible feats, so say the physicists.

Swishing Sounds?

For instance, one of the most common current beliefs is that the aurora borealis makes crackling or swishing noises, and that by applying the proper formula it can be engaged in conversation. Another is that the aurora often comes down to earth, obscuring mountaintops and weaving in and out of bushes, and that an observer with the correct finesse can draw the lights away from or toward him/her at will.

These two alleged attributes of the northern lights are likely to become springboards for losing friendships, as there are few long-time residents of the Northland who have not heard the sounds of the aurora or seen it at eye level.

Reports abound that the aurora whistles, crackles, swishes, snaps or howls. Although scientists have not completely discounted the possibility that noises occur as a result of an aurora, they usually attribute the sounds to ice, snow, breathing or hallucinary disturbances. Dr. Sydney Chapman of Queen's College, Oxford, one of the world's authorities on the aurora, once told a group of scientists meeting at McKinley Park that he was not yet ready to discount entirely the many reports of sounds during displays of the northern lights.

Early Eskimos took for granted that the northern lights made noises, usually ascribing them to the spirits that resided in the lights. When the Eskimos of eastern Greenland heard the lights swishing, they said they were the spirits of children whirling and twisting in their games and dances. Eskimos of northern Canada believed that the whistling and crackling sounds were the footsteps of departed souls tramping about on the snows of heaven.

The Eskimos around Ungava Bay in Canada could hear the spirits speaking to them in a whistling kind of voice, which they took pains to answer in a similar voice. These particular spirits they thought were intermediaries between the living and the dead.

The Eskimos of Western Alaska say today that things are not the same as they used to be, because in the early days the northern lights howled a great deal more than they do now.

Northern Lights Come Close

Although scientists have been reluctant to deny auroral sounds completely, they do deny that northern lights come so close to the earth as to engulf a mountaintop. The aurora never has been measured closer to the earth than 35 miles, and there is doubt whether it actually gets that close. But dozens of lonely northerners have watched their only guest, the aurora, wrap a mountain with its flimsy nets. I have too, scientists to the contrary.

The Brooks Range in Alaska one night in late August was as bright with the aurora as if it were dawn. After two years in the North I had learned to view these dazzling displays with equanimity, but this one differed from others as much as hamburger from a T-bone steak. The aurora had just accomplished the impossible. It had come down to meet the earth just a few miles from me. The top of the highest mountain in the vicinity, almost 9,000 feet of vertical rock, was completely submerged in an auroral gauze.

When it lifted a few seconds later, I assumed I had been the victim of an arctic mirage. But that contrary aurora flickered and dipped over and around the mountaintop for more than 20 minutes, settling now and then with uncanny precision on its pinnacle. What I had witnessed was impossible, according to science. The sensible explanation is that it was reflections of light on skitterish clouds, or some other atmospheric condition. Others who report similar experiences have no doubt seen the same kind of phenomenon.

Eskimo groups as far apart as those of Western Alaska and the Copper Eskimos of Canada were convinced that if a person whistled, the aurora would crackle and then swoop down to earth. The Copper Eskimos also said that if a person spat at them, the various forms of light would run together in the middle and suddenly change form. The Eskimos around Ungava Bay were able to call down the aurora when it talked to them in its whistling voice.

Various ways of talking to the aurora are still popular in Alaska. Several of my Alaska friends have tried to convince me that a human being can make the aurora dart or waver to the earth with a whistle. When I said frankly I did not believe it, one friend volunteered to demonstrate the technique of maneuvering the lights. After a 15-minute whistling demonstration, he turned to me and asked, "You didn't believe it could be done, did you?"

I felt somewhat like the man watching a fisherman who said as he reeled in a sunken boot, "This hook really works, doesn't it!"

My conclusion was that it was a draw. Half of the whistles seemed to draw the aurora nearer to the ground, but half seemed to chase it farther into the sky. My friend used his 50% success as proof that it could be done, and I used his 50% failure as proof that it couldn't. Both of us emerged with original opinions intact.

A generation ago, the Indians living on the Koyukuk River in Alaska thought that whistling was too subtle. They began beating on metal pans to catch the attention of the aurora.

This calling of the aurora implies that it was considered a friendly being or phenomenon. Generally it was, although it is strange that something so flashing and mysterious as the aurora was not more often an object of dread.

Small wonder, though, that the early inhabitants of North America credited strange powers to the northern lights, or that modern people sometimes report unlikely incidents.

In the northern latitudes a display of the aurora begins inauspiciously, with not even a hint of the drama to come in the steady iridescent haze that forms on the horizon early in the evening. The sky, dark and so far away, gradually begins to open up as the haze settles into a comfortable glow, warming the sky above and gradually widening its expanse to immeasurable depths and widths. Then, almost without prelude, the sky moves.

Spectacular Display

Streaks of light toss about with abandon. Gargantuan, ghostlike arms, chasing and darting, appearing and disappearing spontaneously, writhe across the upper sky. Suddenly, for a second, all light melts away and the sky is almost dead with darkness. But just

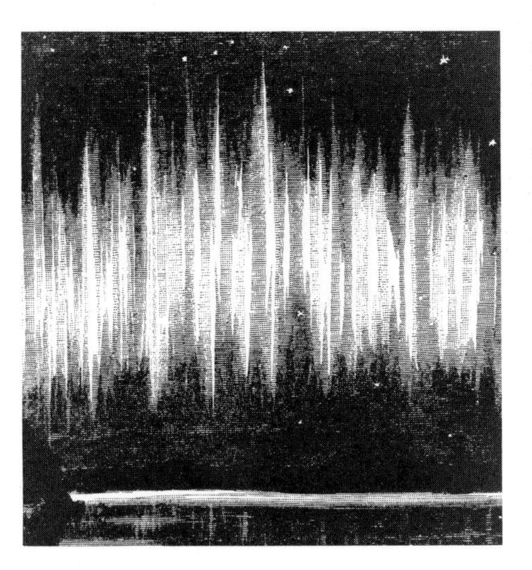

An auroral curtain with a ray structure, sketched by C.F. Hall during his expedition in search of Sir John Franklin in northern Canada. (From Life with the Esquimaux: A Narrative of Arctic Experience in Search of Sir John Franklin's Expeditions *by C.F. Hall, 1864, courtesy of S.-I. Akasofu)*

as quickly the lights blossom again in pulsating waves and arcs, and then, as if to test the credulity of man, giant draperies of light wash in quickly undulating movements across the whole heavens, sometimes stabbing the ends of their folds toward the earth, dripping with the green of grass or the red of blood.

In spite of the flamboyancy of the display, there are only a few recorded instances in which it was presumed to be dangerous or disturbing, and most of these come from the more southerly regions where displays are much less active and much less frequently seen.

Rarity alone was reason enough that they should be viewed with more apprehension. The Eskimos living in the high frequency belt south of the magnetic north pole can see the lights one day out of every three, while in central Mexico and the Mediterranean countries they are usually seen only once in 10 years, 5 times a year in San Francisco, 25 times in Chicago or Seattle, and 100 times in Edmonton, Alberta. Residents of Fairbanks and Nome can see them almost 200 days a year.

Aurora Considered Evil

As far as is known, the Point Barrow Eskimos were the only Eskimo group who considered the aurora an evil thing. In the past they carried knives to keep it away from them.

The Fox Indians, who lived in Wisconsin before they moved to Iowa, regarded the lights as an ill omen of war and pestilence, because the lights to them were the ghosts of their slain enemies who, restless for revenge, tried to rise up again.

Both the Eyak and Tlingit Indians of Southeastern Alaska believed someone would be killed when northern lights played, the Tlingits considering them a sure sign of approaching battle.

The Creek, who lived in the Georgia-Alabama area, and the Cheyenne Indians of Wyoming and Colorado said that the appearance of an aurora meant the weather would change for the worse. The Penobscot Indians of Maine credited the aurora with bringing a windy day. Not only that, but if the lights flickered during the display, the wind would blow strong and steady; if they were of a still and quiet nature, the wind would be squally.

A 19th century saying from Massachusetts went, "South wind and storm will come within 48 hours after northern lights." Scientists have not yet denied that weather changes may be caused by an extremely intense aurora. Certain phenomena such as the expansion of the upper atmosphere may very well upset the lower atmosphere where the weather originates.

To see and hear the aurora was one thing for America's aborigines. To explain it was another. Unequipped with electronic devices and cameras, they had to substitute imagination for research. Explanations of auroral light fell into one of two classes: the aurora originating as fire, or the aurora caused by something alive.

In spite of the fact that the northern lights are often flashing and flamelike, there are fewer beliefs concerning fire than one might expect. The Makah Indians of Washington State thought the lights were fires in the Far North, over which a tribe of dwarfs, half the length of a canoe paddle and so strong they caught whales with their hands, boiled blubber.

The Mandan of North Dakota explained the northern lights as fires over which the great medicine men and warriors of northern nations simmered their dead enemies in enormous pots. The Menominee Indians of Wisconsin regarded the lights as torches used by great, friendly giants in the North, to spear fish at night.

One of the most beautiful beliefs about the aurora is found in an Algonquin myth. When Nanahbozho, the Algonquin creator of the earth, had finished his task of the creation, he traveled to the North, where he remained. He built large fires, of which the northern lights are the reflections, to remind his people that he still thinks of them.

Another tale in which the northern lights are reminders of past deeds is that about Ithenhiela, a Canadian Dogrib Indian character. This story is a variation of the world-wide Magic Flight tale, in which a pursuer is thwarted by the pursued who creates, with magical means, obstacles from very small and often insignificant objects.

The hero of this Dogrib tale flees on a very talented caribou, who, knowing all the tricks, slows down their relentless pursuer by creating hills from a clod of earth, muskeg from a piece of moss, a forest from a tree branch and, finally, the Rocky Mountains from a stone. At the end of his journey Ithenhiela is carried to the sky after pulling a forbidden arrow from a tree. He has lived there ever since, and the northern lights are his fingers moving about.

Spirits of the Dead

Of all the Eskimo groups that have been studied, only one, the Labrador Eskimos, thought that the aurora was caused by fire. The lights to them were torches held by spirits who were seeking to help the souls of the newly dead over the chasm separating the world from the afterworld.

Most of the Eskimo groups from Siberia to Greenland visualized the northern lights as spirits of the dead playing ball with a walrus head or skull. The Eskimos of Nunivak Island had the opposite idea, of walrus spirits playing with a human skull. Even on earth, one of the greatest pleasures for an Eskimo was a ball game, and to look forward to it in the unlimited leisure of heaven was most inviting. This ball-playing heaven, however, was open usually to only those who had died through violent means, so one's ethics in life did not necessarily determine where he would spend eternity.

The use of a whole walrus head with its many pounds of skin suggests a decidedly slow game of ball. Anything could happen in heaven, however, and the lack of gravity in the region of the northern lights probably put some speed into the game. The Siberian Chukchi, a group of people similar in many ways to the Eskimos, said that the head roared while in motion. Not only that, but it would strike with its tusks at anyone who tried to catch it. The walrus skull used by the Hudson Bay Eskimo spirits demonstrated its enjoyment of the game by chattering.

The Salteaux Indians of eastern Canada and the Kwakiutl and Tlingit of Southeastern Alaska interpreted the northern lights as the dancing of human spirits. The Tlingit spirits were particularly happy because they knew that someone who would be killed would come to join them. The Eskimos who lived on the lower Yukon River believed that the aurora was the dance of animal spirits, especially those of deer, seals, salmon and beluga (white whales).

A northern explorer of the late 1700's, Samuel Hearne, reported that to the Chippewa Indians located in central Canada, a bright aurora meant that there were many deer in the sky. The idea evidently originated from the fact that if one strokes the hair of a deer energetically in the dark, he sees sparks.

Spirits of Children

The east Greenland Eskimos thought that the northern lights were the spirits of children who died at birth. The dancing of the children round and round caused the continually moving streamers and draperies of the aurora. When single rays darted out in a horizontal direction, it meant that resident children were running toward arriving children to knock them down.

Occasionally, forms or movements of the aurora were explained in other ways. The Chukchi, for instance, said that the changeable rays were the dead running about in their ball game and the Bering Strait Eskimos said the swaying movement was the struggle of the players. The Iroquois Indians explained the movement as the rising and falling of the sky at a point on earth where souls crossed over into the Land of Souls.

The northern lights also were a source of power to early peoples. Kazingnuk, a man of Little Diomede Island, told me that five generations ago a man of Big Diomede had the aurora borealis as his helping spirit. When this man, Aneuna, was a child, the lights suggested to him the best way to conduct his life. Eventually he became the most influential man in the village, the Eskimo equivalent to a chief. Subsequent persons with the name of Aneuna also received power from the northern lights, but to a diminishing degree. To a Nome woman who is now the fifth generation to bear the name, the power is practically useless.

In this Eskimo area, good persons could become spirits in the northern lights when they died. Aneuna reported that these spirits had continuous happiness and never were thirsty nor hungry. The howling and whistling of the lights are their expressions of happiness.

Dr. Margaret Lantis, who has done extensive research among the Eskimos of Alaska, told me a Nunivak Island tale relating how a poor orphan boy was mysteriously transported to the land of walruses, or the northern lights. When he returned he became a successful medicine man with a walrus-spirit helper.

Eskimo shamans in western Greenland asked the aurora, which represented the souls of the dead, to prescribe for the recovery of the sick. In recent years, Kodiak Islanders regarded the aurora as helpful in curing heart ailments. For example, one boy said that when he had heart trouble, his mother sent for a woman who held him up to the northern lights and then pulled something out of his chest.

But no matter what conclusions were drawn about the lights, no matter what new knowledge becomes ours through scientific investigation, they remain one of the earth's most spectacular sights. The long, dark winters of the Far North will always have their compensating beauty — an ever-changing visual symphony of light and color, the aurora borealis.

Books written by explorers, adventurers and travelers in polar regions often contain their delightful descriptions of the aurora. (From Up in Alaska, *by Esther Birdsall Darling, 1912, courtesy of S.-I. Akasofu)*

There's Magic in the Arctic

By E. L. Keithahn
Reprinted from
The ALASKA SPORTSMAN®
July 1942

The northern lights have seen strange sights. . . .

Thus the poet, Robert Service, introduced one of his poetic tales. However, there are northern lights which are themselves stranger than any imaginary phenomena that the most fertile fictionist could invent. Naturally, mere words or pictures cannot begin to convey an adequate impression of the aurora borealis at its best. Were I to state that on these rare occasions one can "feel" and "hear" and "smell" the aurora, anybody upon reference to a good encyclopedia could refute me instantly. Yet that is exactly the case.

On those arctic nights when the common auroral arch seems to expand, break up and dance overhead in myriad flaming colors, one experiences a mild electrical shock about the ears and hair that is easily distinguished from the nipping of the frost. The air becomes heavily charged with ozone which penetrates the nostrils like chlorine and suggests the smell of blood at a fresh kill. A muffled, swishing sound accompanies these displays. Small Eskimo boys band together and frolic about on the snow attempting to imitate the ominous sounds from overhead. It would be difficult, indeed, to convince these small boys that they weren't hearing anything.

Science places the aurora at approximately 50 to 200 miles above the earth's surface and accounts for it as electrical discharges in oxygen and nitrogen. No doubt the muffled sounds heard in the silent arctic night are actually reverberations of tremendous claps of auroral thunder many miles above the earth.

The common aurora seen night after night in the Seward Peninsula country is a low arch on the northern horizon similar to a rainbow, yet lacking color. Sometimes one arch appears above the other like a double rainbow. Rarely one sees a smaller arch to the left of the main one.

There seems to be a relationship between the aurora and the weather or, strictly speaking, the winds. If it is calm and very cold, the arch is low. As the wind comes up the arch rises, and if it approaches a gale the arch seems to break up at times. The fragments drift about overhead like gigantic draperies. These usually appear as waving sheets of white light, but on occasions color appears in an area, becomes intense, then seems to flit along the entire expanse of the drapery, dying out at the end. These colors rival the brightest rainbow and sometimes the heavens overhead are literally alive with rainbow fragments, dancing, flashing up, then dying out, only to burst out in a spasm of color in some other sector.

Probably the next most interesting and unbelievable phenomenon of the Far North is the mirage. To an enlightened person who has learned the mechanics of mirage they are always a source of wonder and awe. But consider the Eskimo who knows nothing of physics or the laws of optics yet lives in an area that has more mirages than the Sahara.

A cheechako in these northern regions gets the thrill of a lifetime after being frozen in for nine months when he sees the ice breaking up and hears the cry, "Tray-may oomiak-puk kah!" come booming along the coast. It is the signal that the first boat has been sighted, though it might not appear for several days. The call has been relayed from hunting camp to camp for miles along the coast. But when it does show up he can't be sure it's really there!

I recall seeing a beautiful, four-masted schooner appear one morning about two miles offshore in a sea littered with floe ice. And then, right before my eyes, it simply faded out. This kept on intermittently for four days; then the real ship did pull into the roadstead, dropped its hook, and the natives went out in their *umiaks* to barter. When the trading was completed we saw that same ship sail proudly away to the north on the very tip of its masts! It was another antic of mirage.

It is not uncommon for the Eskimos to see the whole Siberian coast suddenly rise out of the sea and thrust itself up like a huge, perpendicular wall thousands of feet high, yet so clear you can image seeing people on its rim. But the supreme hokum is on their own side of the Arctic where there are rugged mountains. These peaks, the Kougaroks, too low and distant to be visible from the coast normally, often rise up like lofty sawtooth ranges with needlelike points. These points then appear to broaden, gradually forming a slender bridge that eventually connects the peaks together, forming an arcaded wall. The walls of the arcades continue to lengthen, then mushroom out until you see a second range of mountains identically like the first except that it is inverted, balancing on the peaks of the initial range. This builds up until at times the mountains are stacked up four ranges high, the alternate ones inverted. The whole spectacle is moving, changing continuously, a blurred unreality that actually looks real.

In the mirage, nature plays her tricks with atmospheric mirrors in a manner similar to stage magicians but on a grander scale than the most skillful magician of all time. Take, for instance, the nights when two moons come up. They are joined together one above the other, the lower one being half beneath the horizon and both blood red. For a while one gets the impression of the head and shoulders of a curious red-hot god of another universe having a look-see at our frozen panorama. Then it stretches its neck inquiringly and finally the head pulls free and floats upward, while the red shoulders recede beneath the rim of the earth.

That is only one of the moon's weird antics. Some nights it comes up red as fire and pulled out of shape like a Japanese lantern or a watermelon on end, then sliced horizontally with the slices slightly apart and evenly spaced. Sometimes it is merely flattened a trifle but with the corners absent. Always the moon in mirage is fiery red. When it rises above the mirage level it assumes its pale color and plays no further tricks.

Sunrises are also affected by the arctic magic. When there is a sunrise in the South, and I mean "South," there is an imitation of it in the North. But as the true sun rises above the horizon its spurious companion fades out. The sunsets are likewise double affairs. One sees light pillars even when there is no sun yet. They appear as a broad band of light reaching perpendicularly into the heavens like a gigantic searchlight. Sundogs, due to ice particles in the atmosphere, are common when the sun comes back in the spring, often eight of them appearing around the sun.

There will be the sun in the center, then above, below and at both sides run four pillars of light like the spokes of a wheel until they meet a ring of light. Where they meet will be four of the sundogs. Then the spokes of light will run out to another halo and there will be four mock suns. This double cartwheel is of tremendous proportions, appearing to occupy about one-sixth of the heavens.

Then there are the frost quakes. It appears the earth can stand just so much freezing, then, due to expansion, something has got to give. As the earth cracks there is a local earthquake, giving the occupants of an igloo directly over it quite a scare, yet not even disturbing their next door neighbor. The sea ice also cracks with a thundering sound like distant cannonading. Small ponds freeze solidly, rise up in the center and erupt. These ponds, frozen convex, can reflect the sun like a large mirror and can be seen for miles when the sun strikes them right.

The arctic will-o-the-wisp is a ball of fuzzy light about the size of a small haystack that silently plunks down in front of you some nights when you are walking along, disappears like magic, appears perhaps behind you, glows with a cold, white light, then bounces away. Now visible, now invisible, it appears and reappears as if it were too hot to alight and not sure it existed anyway. It's just another one of those unexplainable things one encounters in the frozen North.

Other arctic residents have their tales of unbelievable phenomena and the native Eskimos have all had similar experiences which they tell on the long winter evenings. I shall report only one instance since it has a direct bearing on an experience of my own. An Eskimo hunter in the winter of 1924 was set adrift on the ice pack and drifted 600 miles south before he regained the land. It was in the dark days and he was in continual peril of stepping into a crack or an overflow, for in a case like this a wet foot would have cost his life. Questioned about his adventure and miraculous escape the hunter stated that always before him he had seen a narrow white path. He followed this and thus avoided all danger and obstructions such as pressure ridges. Never once was there anything that so much as caused him to stumble on his 600-mile trip in total darkness.

Sometime later an Eskimo father came to my door about midnight asking for medical help for his baby who was seriously ill. I promised to go as soon as I dressed, and he left. When I went out it was pitch black and I had no light. I knew the general direction to his igloo and started off to it, a distance of perhaps half a mile. It was then that as I ran into the blackness I suddenly became aware of a narrow white path before me. It seemed almost to glow softly, and I am not sure that it didn't. Where first I stumbled along gingerly, I now set out full speed with perfect confidence. Some moments later when the unaccountable path abruptly came to an end, I found myself facing the door of the Eskimo father whose sick child needed help.

Snow on the ground reflects the great red aurora of February 1958, photographed in Canada. (E.E. Budzinski)

The Aurora and Polar Explorers

Norwegian Fridtjof Nansen was a respected polar explorer, scientist and statesman at the turn of the century. An illustration of his wife and child hangs on the wall behind him in this sketch. (From Farthest North, *by F. Nansen, 1897, courtesy of S.-I. Akasofu)*

After the great voyages of exploration by such men as Vasco da Gama and Magellan, the polar regions became the remaining frontier for explorers, adventurers and travelers. Many of them were astounded by the beauty of the aurora. Their encounters with the phenomenon were described prolifically in their memoirs, narratives and ship's logs. Among these explorers were Fridtjof Nansen, Adolf E. Nordenskiold, Sir William E. Parry, Sir John Franklin, Elisha Kent Kane, Adolphus W. Greely, Roald Amundsen, Robert F. Scott and Capt. James Cook.

Nansen was, among other accomplishments, a very talented artist and many of his books were illustrated by his own woodcuts and paintings. One woodcut (shown on page 6) depicts a majestic aurora and his ship *Fram*, which was designed by him to withstand the tremendous pressure of the arctic pack ice. During one of his wintering expeditions, Nansen was portrayed by Justin Denzel, an associate, as follows:

> Shortly after midnight, Nansen left the party for a quiet, solitary stroll across the ice. It was a beautiful clear night, with the gaudy streamers of the aurora borealis shifting across the heavens. During his walk, he turned and looked back to see the dark masts and rigging of the *Fram* silhouetted against the pale yellow glow of sky. Behind it, the silken draperies of light were shimmering across the heavens like great pulsating rays of violet sheen, intermingled with pastel shades of pink and green. The very air crackled with the brilliant iridescence, lending an eerie glow to the surrounding landscape.
>
> For almost an hour, he stood there, spellbound by this gorgeous display. In the cold silence, he thought of the long months they had already spent in this vast frozen wasteland and the even longer months that might lie ahead. He thought, too, of Norway and of home, and for a fleeting moment he felt an agonizing pang of regret as he conjured up little Liv and Eva waiting for him on the shore. How long would it be before he saw them again? With a shrug of his shoulders, he pushed the thought from his mind and turned his steps back to the *Fram*, to his merry companions and the warm, comfortable cabins.
>
> Justin F. Denzel, *Adventure North, The Story of Fridtjof Nansen*, London: Abelard-Schuman, 1968, p. 131

After Nansen's attempt to reach the North Pole, the sturdy veteran *Fram* took Roald Amundsen to Antarctica in 1911, where he became the first man to reach the South Pole. Thus, this small ship holds the record for reaching the farthest north and farthest south of any single vessel.

Polar explorers, upon encountering the auroral phenomenon, were often aroused by a great sense of piety. Charles

In this woodcut, Nansen depicts himself strolling on the ice with a triple-curtained form of the aurora overhead. (From Nansen's Nord I Takeheimen, *1911, courtesy of S.-I. Akasofu)*

A very active auroral display over Farmers Loop Road near Fairbanks. (Floyd Damron, reprinted from ALASKA® *magazine)*

A beautiful corona-type aurora, with red emissions from atomic oxygen.
(G.R. Cresswell, Geophysical Institute, University of Alaska)

ABOVE — *Franklin's ships, H.M.S.* Terror *and* Erebus, *trapped in the ice off Cape Felix in northern Canada. The illustration was published in an 1860 book about Franklin's expedition. (Courtesy of S.-I. Akasofu)*
RIGHT — *Britain's tragic hero of antarctic exploration, Capt. Robert Falcon Scott. (From* Scott's Last Expedition, *L. Huxley, 1941, courtesy of S.-I. Akasofu)*
FAR RIGHT — *Sir John Franklin, who led the most tragic polar exploration in history. (Courtesy of S.-I. Akasofu)*

Hall's remark has already been mentioned. Edward Ellis, a 19th century adventurer and author, also exclaimed, "I pity the man who says 'There is no God' or who can look unmoved to the very depths of his soul by such displays of infinite power."

Robert F. Scott, who died in an antarctic blizzard in 1912 after his defeat by Amundsen in the race for the South Pole, described the aurora australis in his daily journal:

> The eastern sky was massed with swaying auroral light, the most vivid and beautiful display that I had ever seen—fold on fold the arches and curtains of vibrating luminosity rose and spread across the sky, to slowly fade and yet again spring to glowing life.
>
> The brighter light seemed to flow, now to mass itself in wreathing folds in one quarter, from which lustrous streamers shot upward, and anon to run in waves through the system of some dimmer figure as if to infuse new life within it.
>
> It is impossible to witness such a beautiful phenomenon without a sense of awe, and yet this sentiment is not inspired by its brilliancy but rather by its delicacy in light and colour, its transparency, and

above all by its tremulous evanescence of form. There is no glittering splendour to dazzle the eye, as has been too often described; rather the appeal is to the imagination by the suggestion of something wholly spiritual, something instinct with a fluttering ethereal life, serenely confident yet restlessly mobile.

One wonders why history does not tell us of 'aurora' worshippers, so easily could the phenomena be considered the manifestation of "god" or "demon." To the little silent group which stood at gaze before such enchantment it seemed profane to return to the mental and physical atmosphere of our house. Finally when I stepped within, I was glad to find that there had been a general movement bedwards, and in the next half-hour the last of the roysterers had succumbed to slumber.

Leonard Huxley, *Scott's Last Expedition, The Personal Journals of Captain R. F. Scott, R.N., C.V.O. on his Journey to the South Pole,* London: John Murray, 1941, p. 257

Incidentally, at present a number of scientific research efforts are being coordinated at the South Pole, the Amundsen-Scott base. A study of the aurora is one such project and the aurora is continuously photographed there.

Out of the countless polar expeditions, the quest for the Northwest Passage, a shortcut from Europe to the Orient across the Canadian polar wilderness, perhaps led most to the kindling of curiosity about the aurora. The passage itself was believed to exist by the opinion that if it was possible for one to pass around the southern tip of South America, one should likewise be able to sail around the northern part of North America. The search was begun by Sir Martin Frobisher who explored the southern part of Baffin Island (Frobisher Bay) in 1576-8. He was followed by many famous explorers, whose names are still familiar in the form of islands, straits and bays in this area, among them Davis, Hudson, Baffin and Parry.

During the middle of the last century, the British Admiralty, becoming impatient at the continued failure to discover the Northwest Passage, decided to send its most experienced arctic explorer, Sir John Franklin, with a crew of 129 officers and men in Her Majesty's Ships *Terror* and *Erebus* to solve this problem once and for all. They sailed in 1845. Franklin's expedition turned out to be one of the most tragic of all polar adventures. The ships were beset by ice in the arctic wastes of McClintock Channel and all members of the party eventually perished after months of hardship and starvation. An old Eskimo woman who had seen them later reported that they "fell down and died as they walked along."

After one of the greatest searches of the 19th century, the remains of some of the members of the lost Franklin expedition were found in 1859 by a party led by Capt. Francis McClintock. After 14 years, the mystery was finally solved. The illustration was published in an 1874 account of the expedition. (Courtesy of S.-I. Akasofu)

In the ensuing years the absence of any trace of Franklin's expedition resulted in the mounting of an intensive search, including a series of expeditions financed by Lady Franklin. Finally, in 1859, a short message was found in a small cairn by Capt. Francis L. McClintock and his search party at King William Island which reads:

> April 25, 1848.—H.M. ships *Terror* and *Erebus* were deserted on the 22nd of April, 5 leagues NNW, of this, having been beset since 12th September, 1846. The officers and crews, consisting of 105 souls, under the command of Captain F.R.M. Crozier, landed here in lat. 69 37 42N., long. 98 41W. Sir John Franklin died on the 11th June, 1847; and the total loss by death in the expedition has been to this date 9 officers and 15 men.
>
> James Fitzjames,
> Captain, H.M.S. *Erebus*.

L. H. Neatby, *In Quest of the Northwest Passage,* New York: Thomas Y. Crowell Company, 1958, p. 176

Numerous memoirs were written by party leaders describing their fruitless search activity. These expeditions, wintering in the Canadian wilderness, witnessed the aurora, astonished by its beauty. It is partly through the extraordinary circumstance of the great search that the beauty of the aurora became widely known to the civilized world at that time.

Sir John Franklin himself was a good observer of the aurora. Although his own memoirs of this ill-fated expedition were never recovered, he left two voluminous reports of his earlier overland expedition from the western shore of Hudson Bay to the Arctic Ocean. In that account he gave the following description:

> For the sake of perspicuity I shall describe the several parts of the Aurora, which I term beams, flashes and arches.
>
> The beams are little conical pencils of light, ranged in parallel lines, with their pointed extremities towards the earth, generally in the direction of the dipping-needle.
>
> The flashes seem to be scattered beams approaching nearer to the earth, because they are similarly shaped and infinitely larger. I have called them flashes, because their appearance is sudden and seldom continues long. When the aurora first becomes visible it is formed like a rainbow, the light of which is faint, and the motion of the beams indistinguishable. It is then in the horizon. As it approaches the zenith it resolves itself into beams which, by a quick undulating motion, project themselves into wreaths, afterwards fading away, and again and again brightening without any visible expansion or contraction of matter. Numerous flashes attend in different parts of the sky.

Sir John Franklin, *Narrative of a Journey to the Shores of the Polar Sea in the Years 1819, 1820, 1821, 1822,* London: J. Murray, 1823, p. 542

Many polar explorers were amazed by the beauty of the aurora borealis. This sketch accurately illustrates a typical curtainlike form of the aurora. (From Recent Polar Voyages *by I.I. Hayes, 1861, courtesy of S.-I. Akasofu)*

Two later arctic explorers pointed to the inadequacy of words to describe the magnificent displays of the aurora.

> Few nights passed without a greater or less display of the Aurora Borealis, that wondrous phenomenon whose existence after more than half a century of research, is yet unaccounted for satisfactorily. Language is vain in the attempt to describe its ever-varying and gorgeous phases; no pen nor pencil can pourtray its fickle hues, its radiance, and its grandeur.
>
> Lt. W. H. Hooper, R.N., *Ten Months Among the Tents of the Tuski, with Incidents of an Arctic Boat Expedition in Search of Sir John Franklin,* London: John Murray, 1853, pp. 384-5

> The aurora of January 21st was wonderful beyond description, and I have no words in which to convey any adequate idea of the beauty and splendor of the scene. It was a continuous change from arch to streamers, from streamers to patches and ribbons, and back again to arches, which covered the entire heavens for part of the time. It lasted for about twenty-two hours, during which at no moment was the phenomena other than vivid and remarkable. At one time there were three perfect arches, which spanned the southwestern sky from horizon to horizon. The most striking and exact simile, perhaps, would be to liken it to a conflagration of surrounding forests as seen at night from a cleared or open space in their centre.
>
> Adolphus W. Greely, *Three Years of Arctic Service, an Account of the Lady Franklin Bay Expedition of 1881-84 and the Attainment of the Farthest North,* New York: Charles Scribner's Son, 1894, pp. 139-40

Although the Northwest Passage was eventually penetrated from east to west by Amundsen in 1903-5, it is still one of the most difficult—and impractical—ocean passages in the world. The largest oil tanker with ice-breaking capabilities constructed in the United States, the S.S.

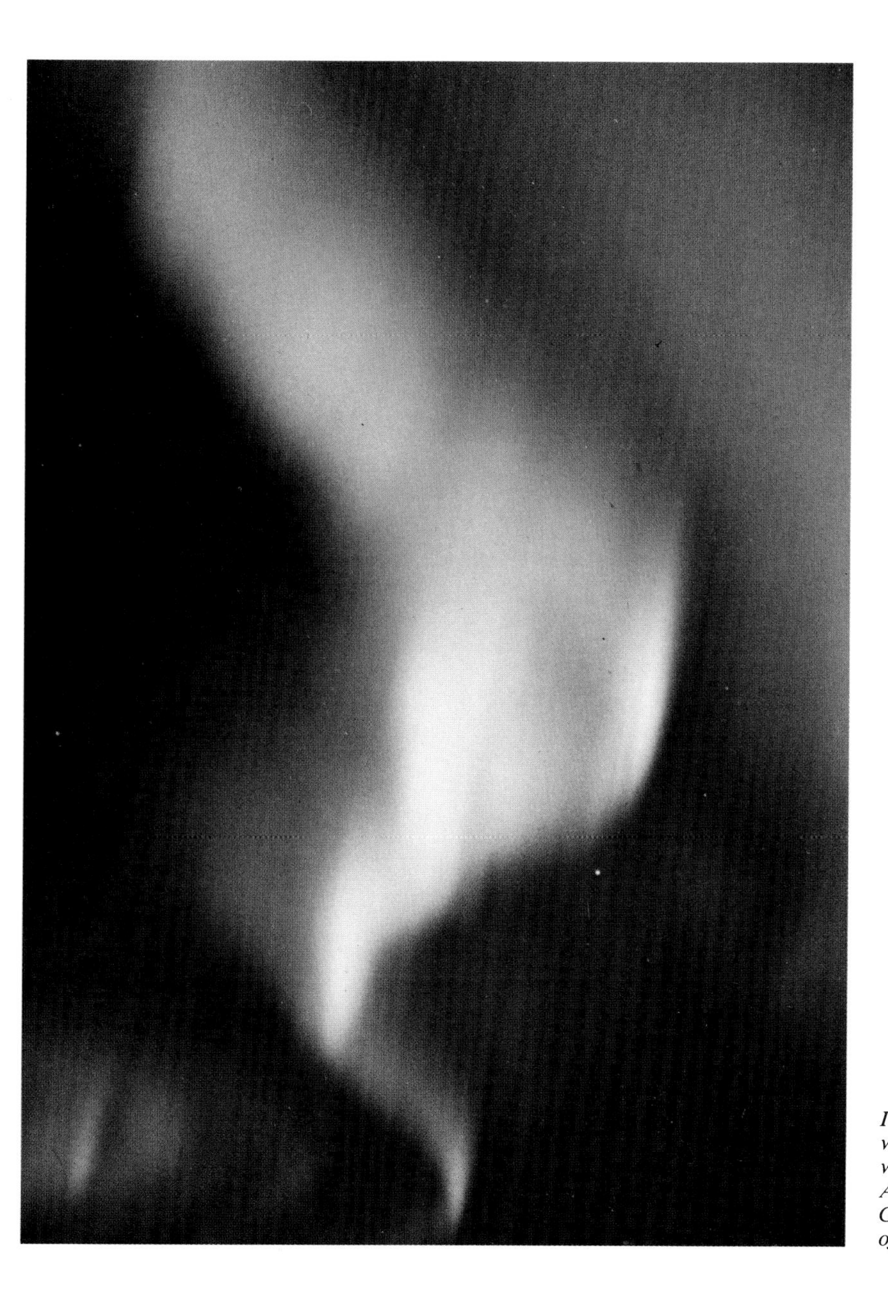

In early Finnish poetry, the aurora was described as a "flaming whirlpool" and the "fire of the Arctic Ocean." (G. Lamprecht, Geophysical Institute, University of Alaska)

"The heavens are beaming with aurora." (From Life with the Esquimaux: A Narrative of Arctic Experience in Search of Sir John Franklin's Expeditions, *by C.F. Hall, 1864, courtesy of S.-I. Akasofu)*

Manhattan, traversed the passage in 1969 in a test of the feasibility of using that route to transport Alaskan oil from Prudhoe Bay. The big ship, encountering great difficulty, barely made it through and the idea was abandoned.

It appears that for polar explorers, the aurora was something quite unique to their life. Thus it is interesting to note the following written by the geologist-explorer, A. E. Nordenskiold, who made the first successful exploration of the arctic passage from Gothenburg, Sweden, to Yokohama, Japan, along the Siberian coast in 1878-9:

> This splendid natural phenomenon . . . plays, though unjustifiably, a great role in imaginative sketches of winter life in the high north, and it is in the popular idea so connected with the ice and snow of the Polar lands, that most of the readers of sketches of Arctic travel would certainly consider it an indefensible omission if the author did not give an account of the aurora as seen from his winter station.

Adolf E. Nordenskiold, *The Voyage of the Vega round Asia and Europe*, London: Macmillan and Co., 1881, p. 36

Another famous explorer and navigator, Captain James Cook, who made three great voyages of discovery to the Pacific, is said to be the first European to witness the aurora in the Southern Hemisphere, the aurora australis. In his log of February 15, 1773, he wrote:

> . . . we had fair weather, and a clear serene sky; and, between midnight and three o'clock in the morning, lights were seen in the heavens, similar to those in the northern hemisphere, known by the name of Aurora Borealis, or Northern Lights; but I never heard of the Aurora Australis being seen before. The officer of the watch observed, that it sometimes broke out in spiral rays, and in a circular form; then its light was very strong, and its appearance beautiful.

James Cook, *A Voyage Towards the South Pole and Round the World*, Vol. 1, London: W. Strahan, T. Cadell, (1777), p. 53

Scenes such as this can be observed when the aurora moves rapidly northward during an intense auroral activity called the auroral substorm. (From The Voyage of the Vivian *by T. W. Knox, 1884, courtesy of S.-I. Akasofu)*

Captain Cook saw the aurora borealis in the Pacific Ocean, on his way from the Sandwich Islands (Hawaii) to Alaska, but curiously, he made no mention of it when he was in Alaskan waters.

As an example of a traveler's description, we may quote from one of the books by the modern (1825-78) Marco Polo, Bayard Taylor:

> All at once an exclamation from Braisted aroused me. I opened my eyes, as I lay in his lap, looked upward, and saw a narrow belt or scarf of silver fire stretching directly across the zenith, with its loose, frayed ends slowly swaying to and fro down the slopes of the sky. Presently it began to waver, bending back and forth, sometimes slowly, sometimes with a quick, springing motion, as if testing its elasticity. Now it took the shape of a bow, now undulated into Hogarth's line of beauty, brightening and fading in its sinuous motion, and finally formed a shepherd's crook, the end of which suddenly began to separate and fall off, as if driven by a strong wind, until the whole belt shot away in long, drifting lines of fiery snow. It then gathered again into a dozen dancing fragments, which alternately advanced and retreated, shot hither and thither, against and across each other, blazed out in yellow and rosy gleams or paled again, playing a thousand fantastic pranks, as if guided by some wild whim.
>
> We lay silent with upturned faces, watching this wonderful spectacle. Suddenly, the scattered lights ran together, as by a common impulse, joined their bright ends, twisted them through each other, and fell in a broad, luminous curtain straight downward through the air until its fringed hem swung apparently but a few yards over our heads. This phenomenon was so unexpected and startling, that for a moment I thought our faces would be touched by the skirts of the glorious drapery. It did not follow the spheric curve of the firmament, but hung plumb from the zenith, falling, apparently, millions of leagues through the air, its folds gathered together among the stars and its embroidery of flame sweeping the earth and shedding a pale, unearthly radiance over the wastes of snow. A moment afterwards and it was again drawn up, parted, waved its flambeaux and shot its lances hither and thither, advancing and retreating as before. Anything so strange, so capricious, so wonderful, so gloriously beautiful, I scarcely hope to see again.
>
> *Prose Writings of Bayard Taylor*, Northern Travel: Norway, Lapland &c., New York: G. P. Putnam, 1864, pp. 63-4

ABOVE — *The shape of the aurora in this 19th century engraving may look very distorted and abstract. This particular form is called the corona, seen when the aurora curtain is located near its zenith. (From* Under the Rays of the Aurora Borealis *by Sophus Tromholt, 1885, courtesy of S.-I. Akasofu)*

RIGHT — *This photograph proves that the above engraving is an accurate rendering of the coronal form. (S.-I. Akasofu)*

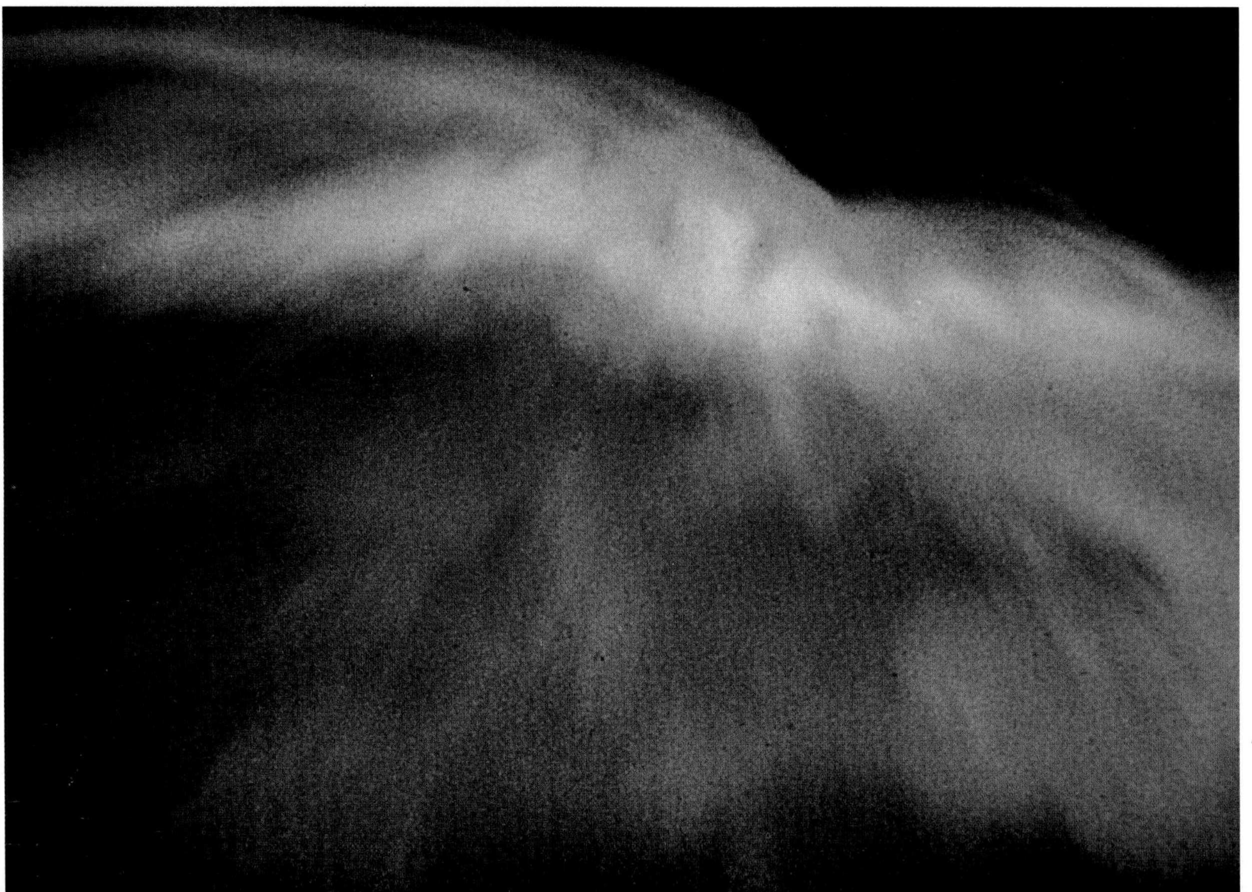

The Aurora and Northern Settlers

Following the exploits of the polar explorers, it is natural that settlers, traders and miners would flock to inhabit the new land. There were those who sought riches, particularly gold and furs, and those who wanted to start a new life with a healthy dose of adventure and the promise of a peaceful existence in the wilderness. Numerous accounts of the aurora were left by these early settlers; it was a phenomenon little understood by them, but frequently mentioned in their notes and letters. Verses were composed, perhaps inscribed by lantern or candlelight in the long darkness of the northern winter, when the cold allowed only a brief step out of the cabin's

An auroral display over Interior Alaska. Early settlers described the aurora in various ways, including "a vast wall of color," "raining streams of brilliants" and "the blaze of Alaska." (S.-I. Akasofu)

Some early settlers imagined a wavy auroral curtain as a graceful snake of electric light. (From Travel and Adventure in the Territory of Alaska *by Frederick Whymper, 1868, courtesy of S.-I. Akasofu)*

warmth to marvel at the wonder in the sky. Naturally, these witnesses viewed the aurora each in his own way, depending on their purpose for being in the Far North. Frederick Whymper wrote:

> Just as we were turning in for the night a fine auroral display in the N.W. was announced, and we all rushed out to witness it from the roof of the tallest building in the Fort. It was not the conventional arch, but a graceful, undulating, every-changing "snake" of electric light; evanescent colours, pale as those of a lunar rainbow, ever and again flitting through it, and long streamers and scintillations moving upward to the bright stars, which distinctly shone through its hazy, ethereal form. The night was beautifully calm and clear, cold, but not intensely so, the thermometer at +16°.
>
> Frederick Whymper, *Travel and Adventure in the Territory of Alaska,* London: John Murrray, 1868, p. 178

Some gold miners thought the aurora was a vapor from a hidden mine. Robert Service presented this view tongue-in-cheek in his poem, *The Ballad of the Northern Lights.*

> *Some say that the Northern Lights are the glare of the*
> *Arctic ice and snow;*
> *And some that it's electricity, and nobody seems*
> *to know.*
> *But I'll tell you now - and if I lie, may my lips be*
> *stricken dumb-*
> *It's a* mine, *a mine of the precious stuff that men*
> *call radium.*
> *It's a million dollars a pound, they say, and there's tons*
> *and tons in sight.*
> *You can see it gleam in a golden stream in the solitudes*
> *of night.*
> *And it's mine, all mine - and say! if you have a hundred*
> *plunks to spare,*
> *I'll let you have the chance of your life, I'll sell you a*
> *quarter share.*
>
> Robert Service, *Collected Poems of Robert Service,* New York: Dodd, Mead and Company, 1940, p. 89

When an auroral curtain is several hundred kilometers north of an observer, it will be seen as an archlike luminosity. Because of the perspective effect, both the western and eastern ends seem to rise from the horizon, an effect well illustrated in this woodcut by Fridtjof Nansen. (Courtesy of S.-I. Akasofu)

During an active display near Fairbanks, the auroral curtain appears to break into groups of rays, crisscrossing in an endless dance in the chilly sky. (G. Lamprecht, Geophysical Institute, University of Alaska)

A gold miner, Stanley Scearce, ruminated on his travels:

That evening as we sat around the campfire, Charlie told me about his country and the meat and furs they had brought to Dawson to trade. Far away on the northern horizon the aurora rose and fell with brilliant multicolors.

"Charlie, how can I get Northern Lights to help me?"

"Mebbe talk to them. They help you be big trader."

"How, Charlie? I want to get gold."

"Best way to get gold is to have store. All 'um gold go to storekeepers. White man dig gold; show man and man who plays cards get 'um; storekeeper all time get that too."

"I wish I could be storekeeper, Charlie!" I felt that this wise old chief could help me. "Would your Great White Spirit help me too?"

"Help all time. You make 'um big talk. He tell you what do."

Stanley Scearce, *Northern Lights to Fields of Gold*, Caldwell, Idaho: The Caxton Printers, Ltd., 1939, p. 98

Another example of the outpouring of verse by the settlers and gold miners in Alaska and the Yukon is from F. B. Camp's book, *Alaska Nuggets*.

I'm a Sour-dough of Alaska
Who knows the Give and Take;
At Dawson, Nome, Iditarod,
I grabbed a handsome stake,
Then mushed it to the Outside
With all my Golden Pokes,
And invested them in business
That throttles me and chokes.

Today while working in my chair,
In office high up in the air,
Surcharged with smoke and other hazes;
My vision played a trick on me,
And set my eyes a roaming free,
Till once again I glimpsed the Wonder Blazes.

I saw the Blaze of Alaska, on the Blaze was a picture dear,
And the steaming haze of the housetops parted till it was clear.
It showed me the Trails now hidden, it pictured the mountains high,
The river sands, where with their hands, men pan for gold till they die.

It revealed the swamps and the forests, where the big moose live and the bear;
Where the ptarmigans nest, on the high hill's crest, where the black fox has his lair.
It painted the bright aurora, that shines with a myriad lights,
As it brightens the snow, where the North winds blow, chill on the winter nights.

Frank B. Camp, *Alaska Nuggets*, Anchorage, Alaska: Alaska Publishing Co., 1922, p. 19

A scientist makes a final check of an all-sky camera during an arctic sunset. (Geophysical Institute, University of Alaska)

Scientists Challenge the Mystery

Like any other field of science, our knowledge of the aurora has come a long way. It may be helpful to an understanding of what is now known about the phenomenon to present the facts in the process of looking at the history of the science.

Pierre Gassendi was said to be the man who introduced the term "aurora" in the way we presently use it. Gassendi, a 17th century scientist, mathematician and philosopher, was also the first to observe a planetary transit, that of Mercury in 1631, predicted by the great astronomer Kepler. The name came from Aurora, the rosy-fingered Goddess of the Dawn in Roman mythology, who was the herald of the rising sun.

A problem that claimed the attention of many early auroral scientists was how to determine the height of the aurora. There was much heated controversy about it, since reported heights ranged from ground level to an altitude of 1,000 km (621 miles). Numerous accounts were published in which observers declared, for example, that the aurora appeared between two houses, or a few thousand yards away, or around the tops of mountains. Even in the *Encyclopedia Britannica* (9th Edition, 1882; 11th Edition, 1910) some of these reports were quoted as reliable. Thus, many scientists sought the cause of the aurora in the lower atmosphere of earth where clouds are formed, below the altitude at which jet aircraft fly today. As we shall see, those authors were confused by the effects of perspective.

One of the first scientists to conclude that the aurora is a phenomenon which appears well above cloud levels was the Frenchman, De Mairan, who wrote the first monograph on the subject in 1754. In fact, he went so far as to infer that the aurora was caused by the impact of an extraterrestrial material on the upper atmosphere.

Two famous British scientists, Henry Cavendish (1731-1810) and John Dalton (1766-1844), were accurate in determining the height of the aurora to be 80-250 kilometers (50-155 miles), while Adam Paulsen, a Danish polar explorer, in 1889 reported it to be 600 meters-67 kilometers (2,000 feet-43 miles). Carl Stormer (1874-1957), a Norwegian physicist, made the most extensive height determination by taking auroral photographs simultaneously from two widely separated points and solving the problem by triangulation. His finding was that the bottom of most of the aurora is 100-

TOP — *Pierre Gassendi, said to have introduced the term "aurora." (Courtesy of S.-I. Akasofu)*
ABOVE — *John Dalton, an outstanding scientist, was intrigued by the aurora and spent considerable time in its study. (The Royal Society of Scotland)*

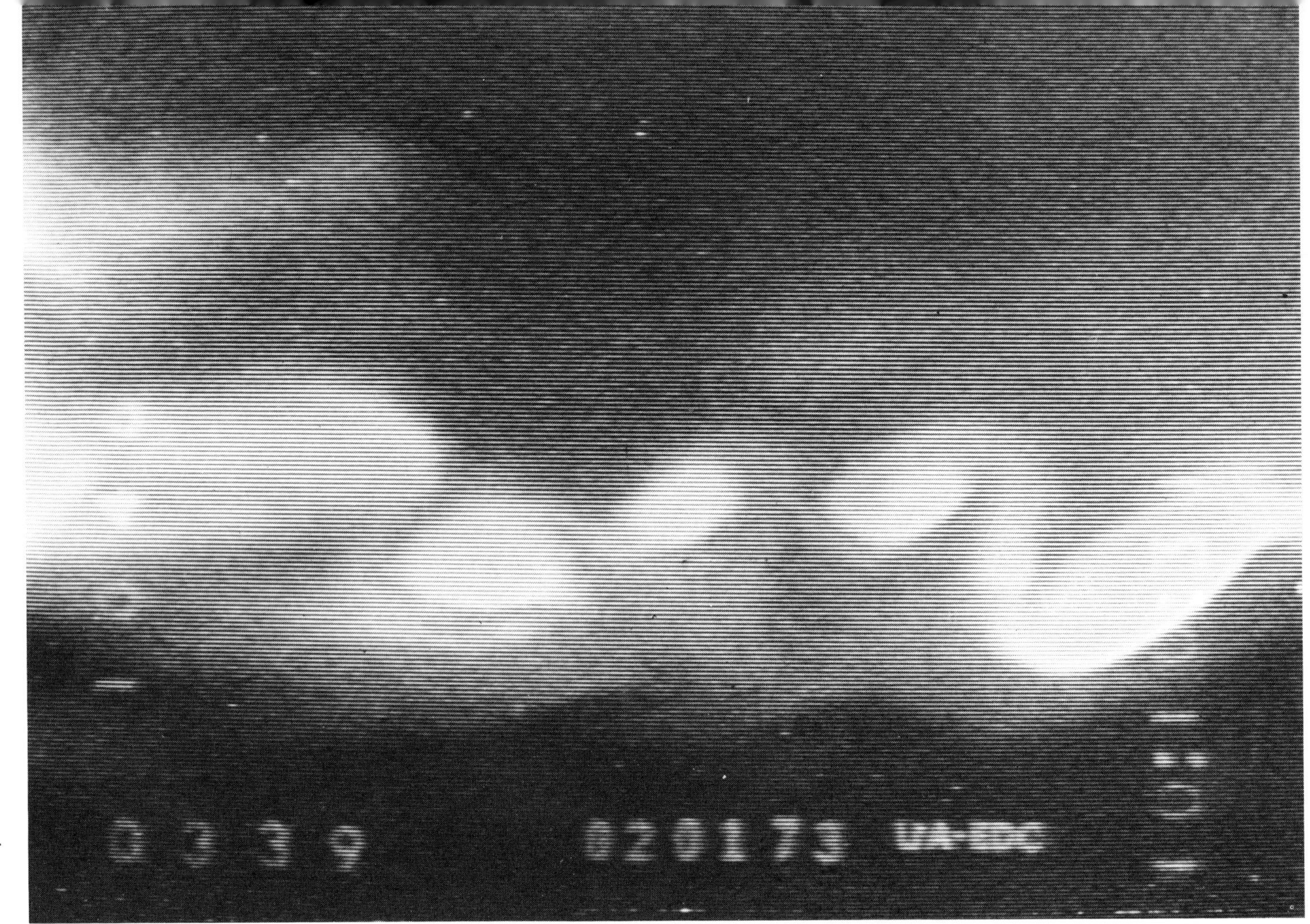

Auroral rays, as seen from the bottom of the curtain by a supersensitive, high-speed television system (about 30 frames per second). Each eddylike structure has a diameter of a few kilometers and will be seen as a "ray" when the auroral curtain is observed from a distance. A Russian cosmonaut who went through active auroras mentioned that he felt as if he were passing through "magnificent columns of light." (T.N. Davis and T.J. Hallinan, Geophysical Institute, University of Alaska)

105 kilometers (62-65 miles), about 10 times higher than the cruising altitude of jet aircraft and the highest clouds. The maximum altitude varies considerably, from a few hundred to more than a thousand kilometers.

Most polar auroras have basically a curtainlike or ribbonlike form. This curtain formation does not, however, hang vertically but inclines slightly southward (the reason for this inclination is given in the next chapter). In its simplest aspect the surface brightness of the aurora (brightest near the bottom) is fairly even in the horizontal direction. As the auroral activity increases, folds develop, the complexity and extent of the folds depending on the degree of activity. When the aurora is slightly activated, nearly vertical striations, called the "ray structure," appear. These are actually very fine folds and can be seen best when a curtainlike form is located a little south of the observer, who can see only the bottom edge of the curtain because of the inclination. The folds also often develop an eddylike structure. Motions of these eddies are frequently quite violent and swift; from north or south of the auroral curtain, these motions are seen as a very rapid horizontal motion of the "rays." As a result, an ordinary movie camera cannot record them properly, even though the sensitivity of films has been considerably improved in recent years. A special TV technique is needed to record such motions. As activation increases, the scale of the

HEIGHT OF THE AURORA: *The altitude of the aurora as compared with Mount Everest, (29,028 feet elevation), a balloon and an aircraft capable of reaching high altitudes.*

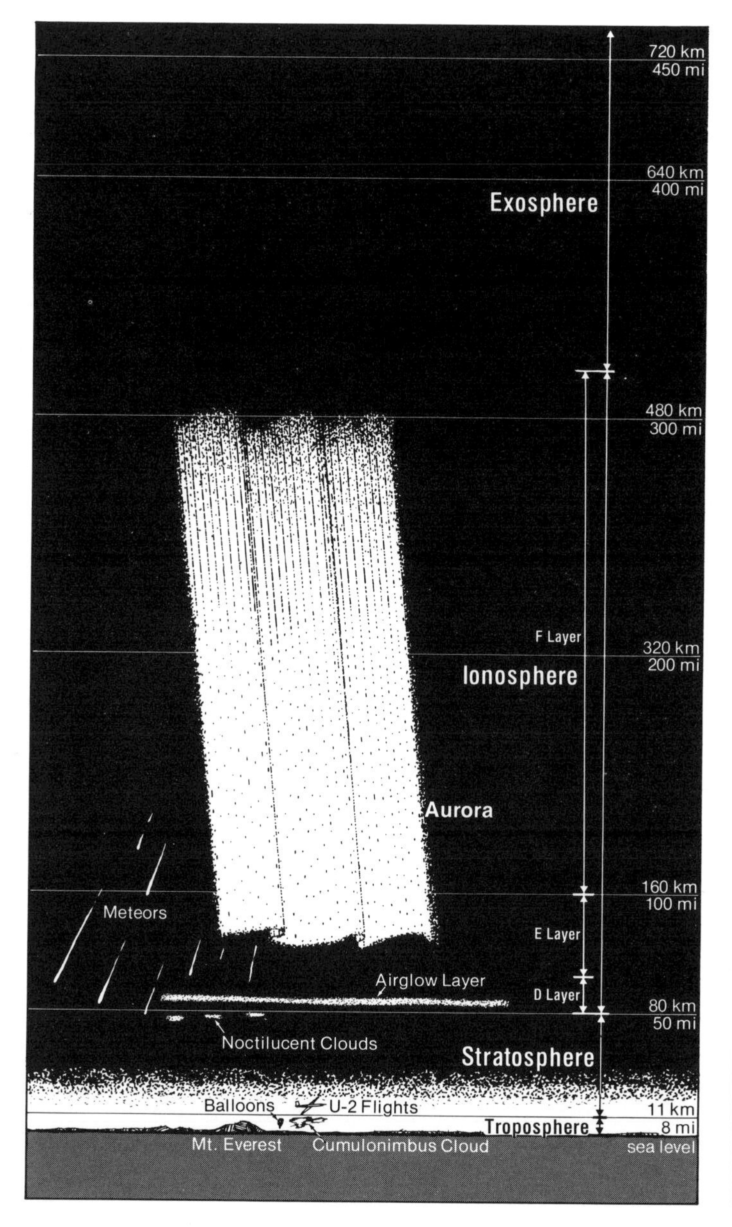

COMMON FORM OF THE AURORA: *The basic curtainlike form develops a variety of folds as it is activated. The scale of folds increases as the degree of activity is enhanced.*

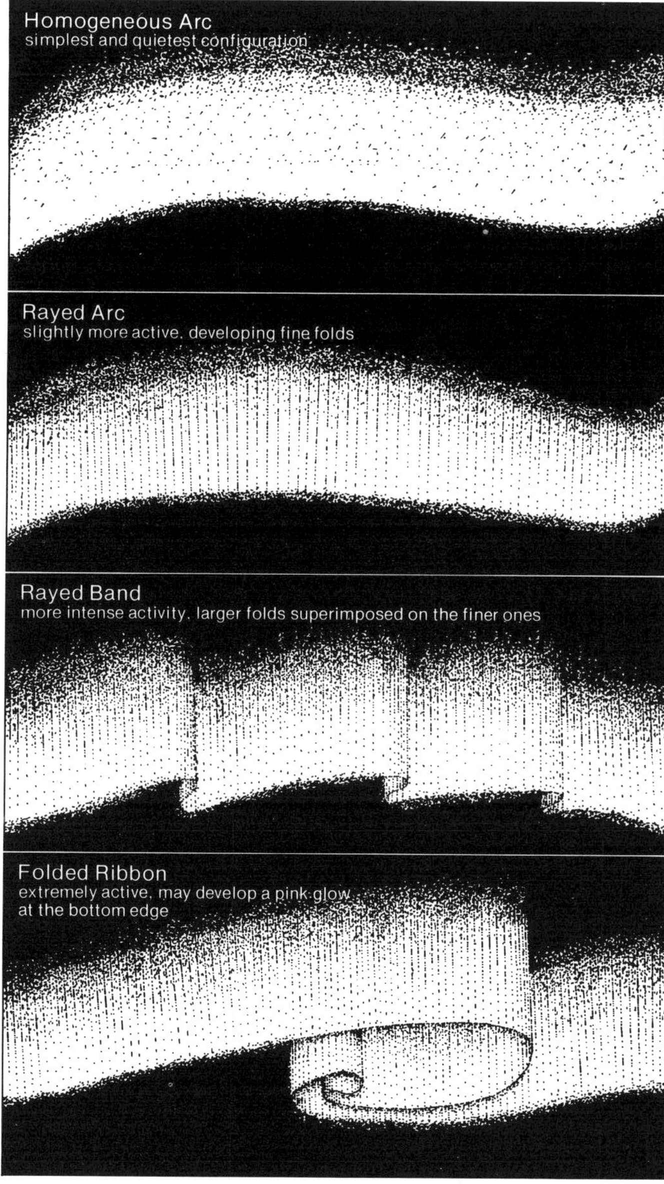

folds increases, developing into a large wavy structure. In its most active state, the auroral curtain develops a loop structure that extends horizontally 100-200 kilometers (62-124 miles), and eventually achieves a curled, drapery form.

Actually, the aurora is a very large-scale phenomenon encircling the entire polar region, but when you view a particular display in the chilly night sky, you see only a very small portion of the curtainlike form. Therefore, it is not apparent to most observers that all bright forms are curtainlike. Obviously, astronauts and cosmonauts, looking down on the entire polar region from high above, are in a much better position than those of us who are earthbound to observe the aurora.

An auroral curtain with a large-scale fold. Fairbanks residents often enjoy such displays during the Christmas season. (M. Grassi)

LEFT — *In his painting of a church at Utsjoki, Finland, artist Harold Moltke captured a rayed band of aurora. (Meteorologisk Institut, Copenhagen, courtesy of S.-I. Akasofu)*
BELOW — *An active auroral curtain sometimes develops a loop structure, as this photograph shows. (S.-I. Akasofu)*

These photographs of the aurora were taken at the University of Alaska, Fairbanks, during the early 1950's. In the left photograph, two auroral curtains lie nearly in parallel. In the right photograph, a looped curtain is seen. (V.P. Hessler, Geophysical Institute, University of Alaska)

A skylab of an earlier day. On August 24, 1804, the two famous physicists Gay-Lussac and Biot ascended in a balloon to 13,120 feet to take samples of the air and measure the earth's magnetic field. The caged bird was carried along to warn the scientists of a possible dangerous lack of oxygen, since it was believed the bird would succumb to the thinner air before the men. (Courtesy of A. Lebeau)

When such a large curtainlike form is seen by an observer who is located a few hundred kilometers south of it, it will be seen as an archlike luminosity across the northern sky. In this situation, its western and eastern "ends" are about 1,000 kilometers (621 miles) away from the observer. One can see such a great distance because the auroral curtain is at an altitude of about 100 kilometers (62 miles). This causes a perspective effect, and both the western and eastern "ends" seem to touch the ground at the horizon, as Nansen's woodcut on page 35 shows.

When the curtainlike form is located near the zenith of

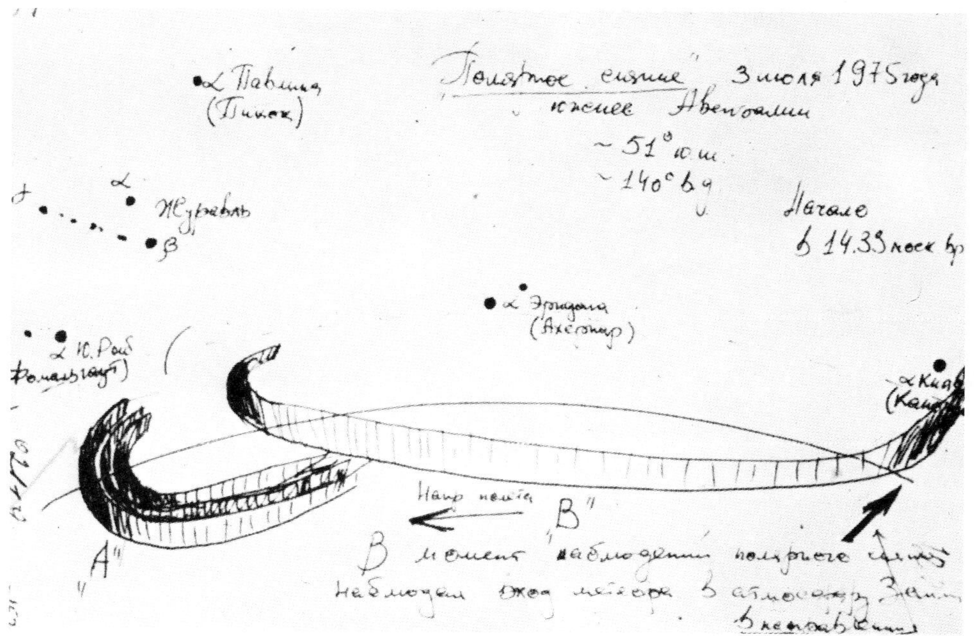

A sketch of the aurora made in 1975 by Russian cosmonaut Vitalii Ivanovich Scvastyanov aboard Salyut-4. *A curtainlike form encircling the polar region is well illustrated. (From* Optical Studies of Atmospheric Emissions, Aurora Borealis, and Noctilucent Clouds Aboard the Orbital Scientific Station "Salyut 4," *1977)*

The aurora, photographed from above by Astronaut O.K. Garriot aboard Skylab. Also clearly seen is the curved edge of the earth and the luminous layer at about 90 kilometers (51 miles), called the airglow layer.

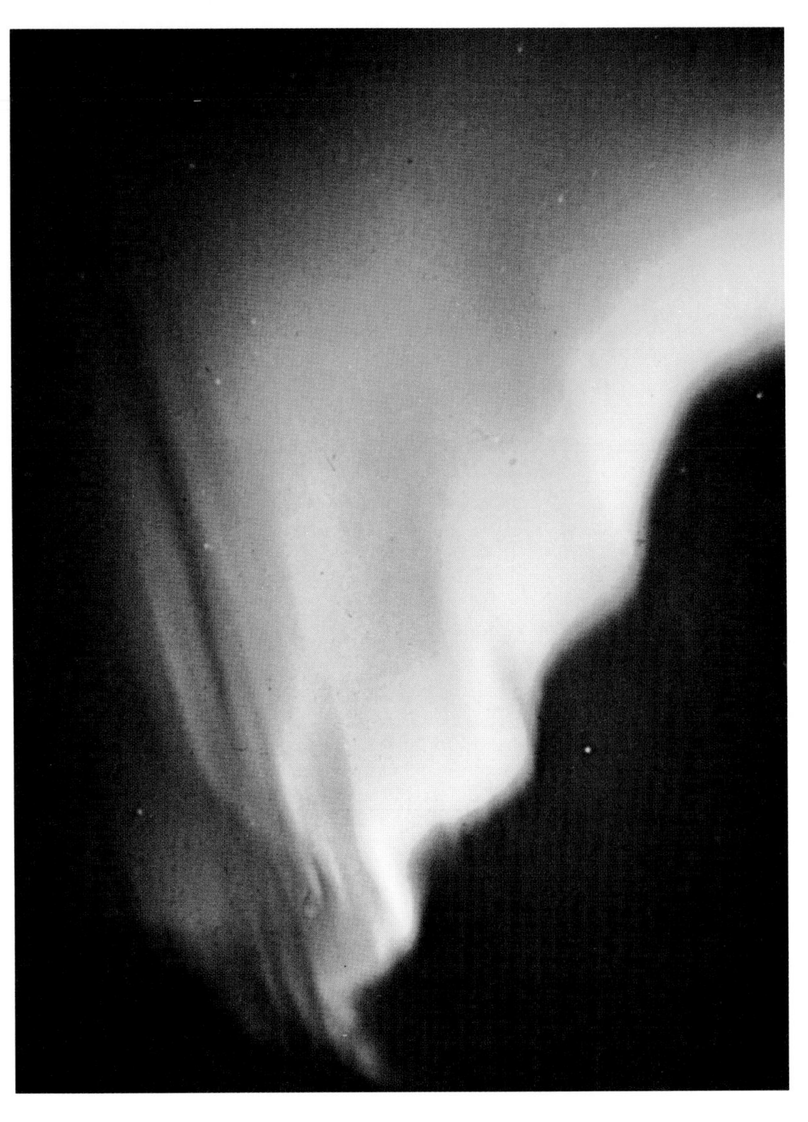

Because of an effect of perspective, when the curtainlike form is located near the observer, the eastern or western end of the curtain appears to rise like smoke from the ground at the horizon. (G. Lamprecht, Geophysical Institute, University of Alaska)

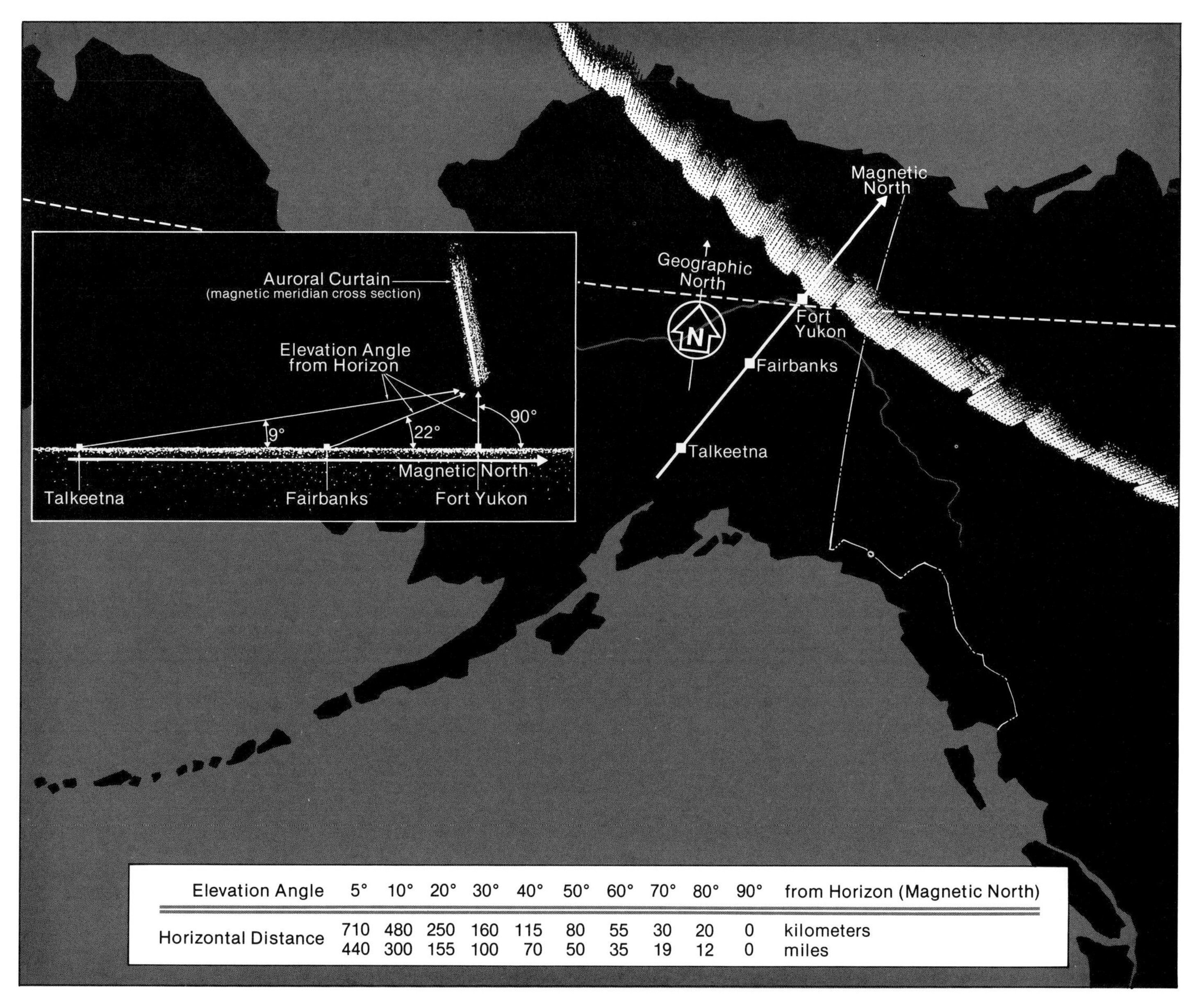

Elevation Angle	5°	10°	20°	30°	40°	50°	60°	70°	80°	90°	from Horizon (Magnetic North)
Horizontal Distance	710	480	250	160	115	80	55	30	20	0	kilometers
	440	300	155	100	70	50	35	19	12	0	miles

VIEWING ANGLES AND HORIZONTAL DISTANCE OF THE AURORA: *If an observer measures the angle between the bottom of an auroral curtain and the magnetic north horizon, he can get the horizontal distance between himself and the aurora. If, for example, the auroral curtain is over Fort Yukon, Alaska, which is 135 miles magnetic north of Fairbanks, the angle of sight from the bottom of the auroral curtain to the magnetic northern horizon is about 22 degrees in Fairbanks. Similarly, it will be seen 9 degrees above the horizon from Talkeetna, which is about 315 miles magnetic south of the aurora.*

an observer, the perspective effect can be recognized more clearly. In this situation, the aurora appears to rise, often like smoke, from the eastern or western horizon. It was such a situation that led some gold miners to believe that the aurora was vapor from a mine, a belief played upon by Robert Service in his poem.

When a very active auroral curtain with large-scale folds is located near the zenith, it is difficult for anyone to

PERSPECTIVE EFFECT AND THE CORONA FORM: *When an auroral arc is slightly south of an observer, he will see the bottom of an auroral curtain. When the aurora is active, showing wavy forms and particularly a drapery or loop form, he will see the corona form, a spectacular display. This is a perspective effect. The observer is seeing a large number of rays which are nearly parallel, but extend at least a few hundred kilometers above him, so that the rays appear to converge. When the* same *aurora is viewed simultaneously by an observer a few hundred kilometers away, it will be seen as a curtainlike form.*

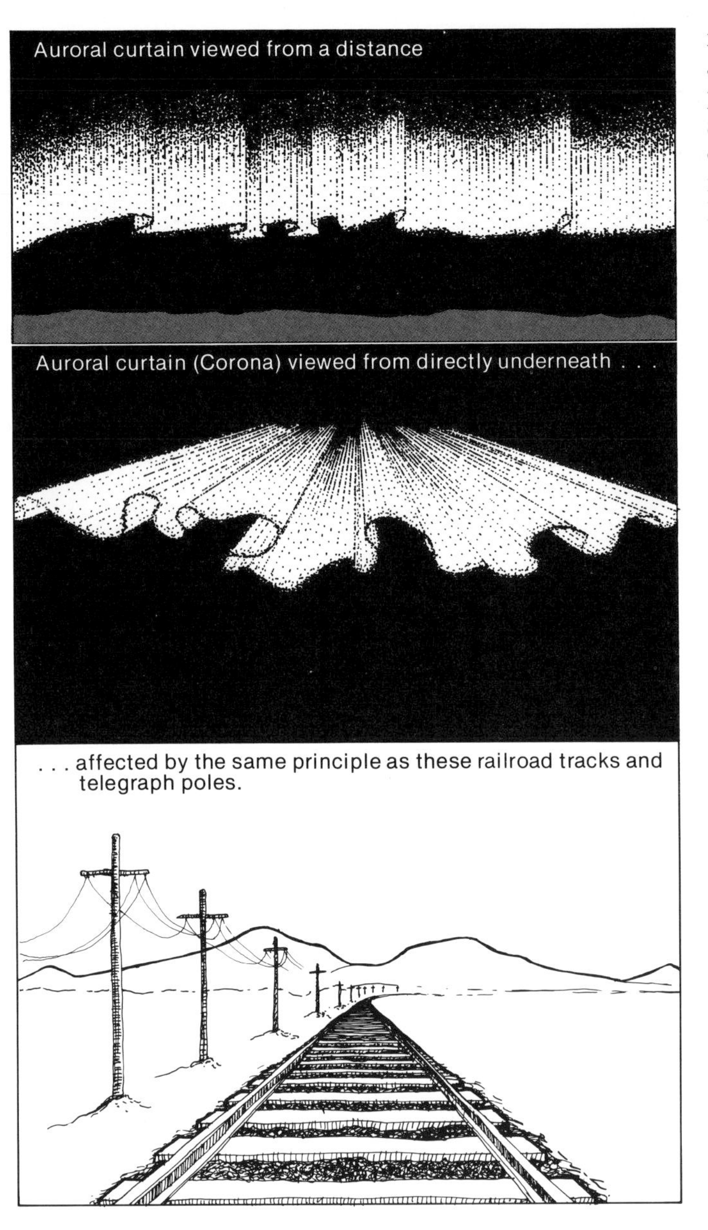

recognize the curtain form and, in fact, it appears to have an entirely different form, called the corona. Again, the perspective effect is playing a role here. The observer is seeing a large number of parallel rays. However, since they extend at least a few hundred kilometers above him, the rays appear to converge at a point in the sky which is located slightly south of the observer (because of the slight southward inclination of the curtain). When the same aurora is viewed simultaneously by an onlooker located a few hundred kilometers north or south of him, that observer will see an ordinary curtainlike form. Therefore, contrary to common belief, the corona is not really a different form of the aurora. Its appearance indicates that an observer is almost directly beneath the auroral curtain. An interesting description of the coronal form is found in an account by the American arctic explorer, Charles Francis Hall (1821-71). Hall, after the discovery of the fate of the Franklin expedition, set out in search of members of that tragic party whom he believed might be living among the Eskimos.

> The heavens are beaming with aurora. The appearance of this phenomenon is quite changed from what it has been. Now, the aurora shoots up in beams scattered over the whole canopy, all tending to meet at zenith. How multitudinous are the scenes presented in one hour by the aurora! This morning the changes are very rapid and magnificent. Casting the eye in one direction, I view the instantaneous flash of the aurora shooting up and spreading out its beautiful rays, gliding this way, then returning, swinging to and fro like the pendulum of a mighty clock. I cast my eyes to another point; there instantaneous changes are going on. I close my eyes for a moment; the scene has changed for another of seemingly greater beauty. In truth, if one were to catch the glowing heavens at each instant now passing, his varied views would number thousands in one hour.

Charles F. Hall, *Arctic Researches and Life Among the Esquimaux: Narrative of an Expedition in Search of Sir John Franklin, in the Years 1860, 1861, and 1862.* London: Sampson, Low, Son and Marston, 1864, p. 151

A pencil sketch by Nansen of the corona form, drawn in December 1894.
(From Farthest North, *F. Nansen, 1897, courtesy of S.-I. Akasofu)*

There is another type of aurora, called the diffuse aurora. It is a diffuse luminosity that appears to be a few hundred kilometers wide, stretching across the sky in an east-west direction. It looks almost like the Milky Way, often slightly brighter than that portion of our galaxy, and it appears a little south of the curtainlike form.

The question of how frequently the aurora can be seen in an average year has been examined by a number of auroral scientists. In 1860, after a painstaking compilation of auroral records from the past, Elias Loomis, a professor at Yale University, produced a map which shows three belts. In one, the aurora is seen 80 nights per year and in the other two, 40 nights. It is amazing how accurate many of the scientific works are by today's standards, despite the fact that the data available at the time must have been scanty by our standards. Loomis's work is a good example of this kind, even though his map has since been greatly refined.

Below is a map produced by E. Harry Vestine, an American auroral scientist, in 1944. The number in each curve indicates the number of nights per year in which the aurora can be seen: "0.1" means 1 night in 10 years; "1," 1 night per year; "243" means 243 nights per year. A narrow belt centered around this "243" curve is called the auroral zone. These numbers are based on a long-term average. Since the appearance of the aurora depends greatly on the sunspot cycle (a period of about 11 years), the numbers are generally higher than those given in the map during the period when the sunspot occurrence is high (e.g., 1979-81; 1990-3).

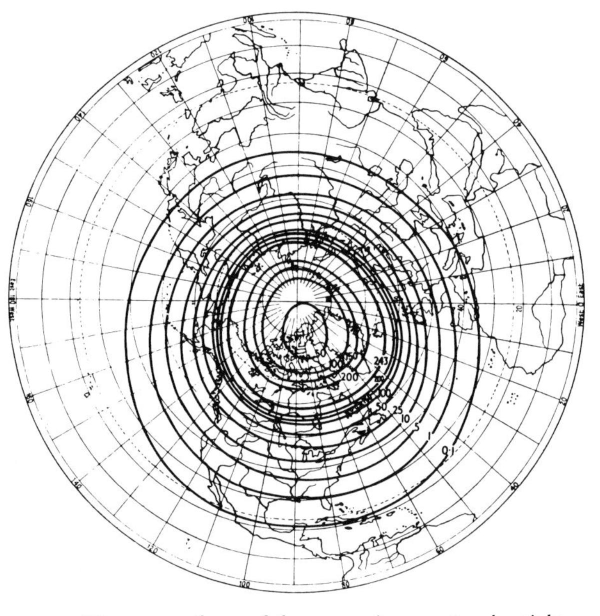

LEFT — *The corona form of the aurora is a spectacular sight. (S.-I. Akasofu)*

ABOVE — *The average frequency of auroral displays (the number of nights per year) determined by E.H. Vestine. (*Terrestrial Magnetism, *Vol. 49, p. 77, 1944, courtesy of S.-I. Akasofu)*

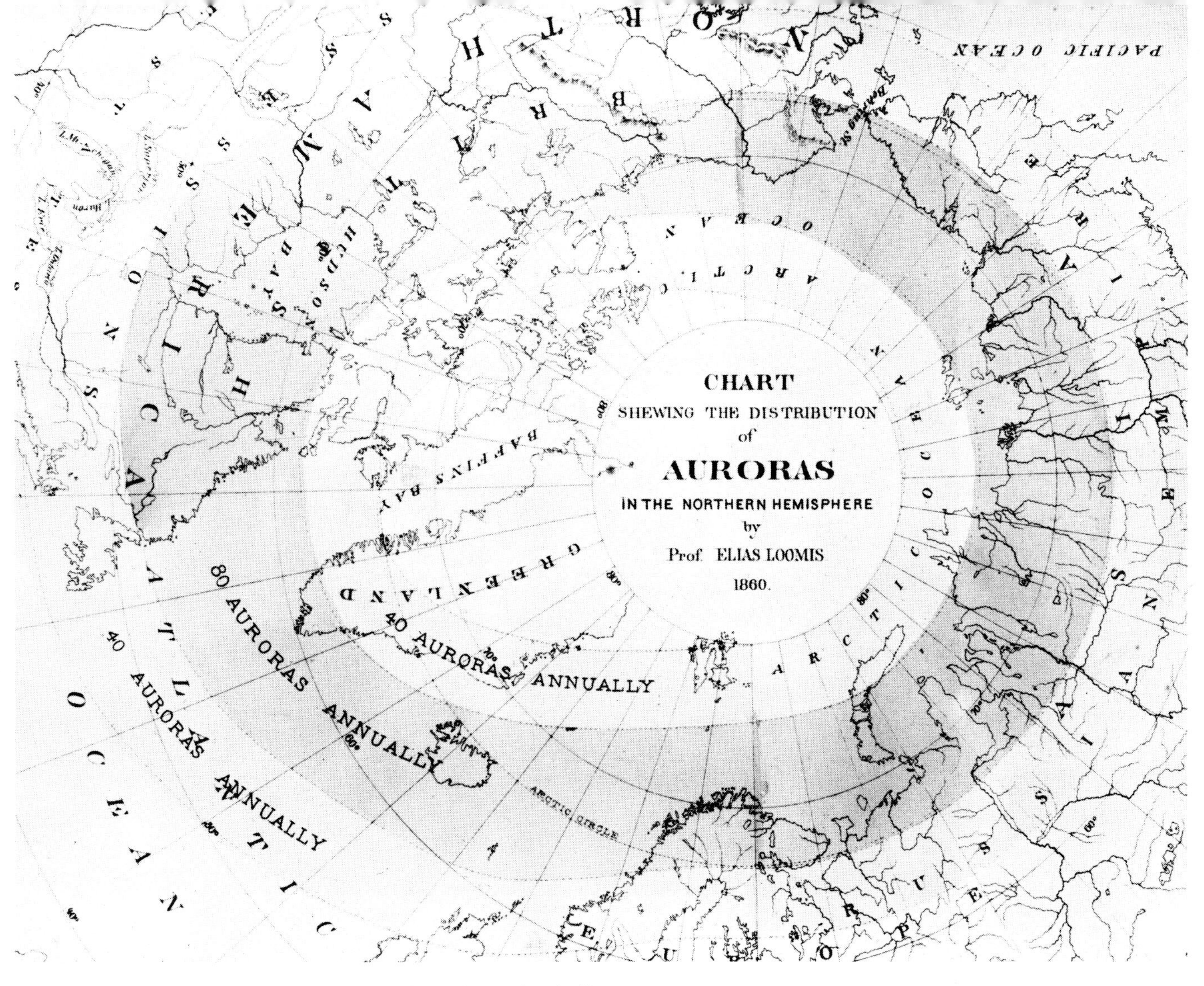

LEFT — *The first chart showing the average annual frequency of occurrence, produced by Elias Loomis in 1860.* (American Journal of Science and Arts, *1860, courtesy of S.-I. Akasofu)*

BELOW — *Elias Loomis, first scientist to notice that the aurora appears along a narrow band surrounding the pole. (Courtesy of Yale University)*

The maps constructed by both Loomis and Vestine indicate that the occurrence of the aurora reaches maximum at a latitude of about 65° and then decreases toward higher latitudes. Robert E. Peary, during his quest of the North Pole observed this tendency.

> Nature herself participated in our Christmas celebrations by providing an aurora of considerable brilliancy. While the races on the ice-foot were in progress, the northern sky was filled with streamers and lances of pale white light. These phenomena of the northern sky are not, contrary to the common belief, especially frequent in these most northerly latitudes. It is always a pity to destroy a pleasant popular illusion; but I have seen auroras of a greater beauty in Maine than I have ever seen beyond the Arctic Circle.
>
> Robert E. Peary, *The North Pole*, London: Hodder and Stoughton, 1910, p. 172

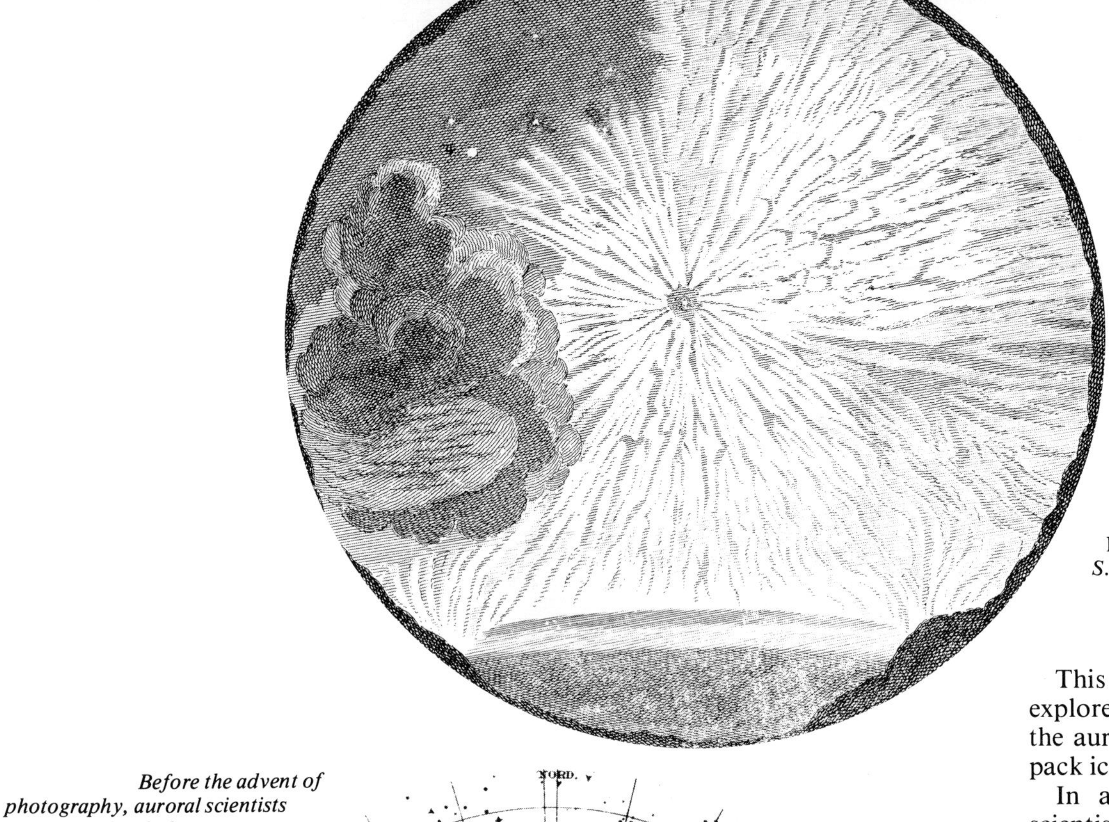

One of the first scientific sketches of the aurora over the whole sky done by De Mairan in 1726. (From Traite Physique et Historique de l'Aurore Boreale *by De Mairan, 1733, courtesy of S.-I. Akasofu)*

Before the advent of photography, auroral scientists had to sketch the aurora. This sketch was made by V. Carlheim-Gyllenskold when he made an observation in Spitzbergen. (Courtesy of S.-I. Akasofu)

This same fact was noted by several other polar explorers, and it led some to conclude (erroneously) that the aurora was prone to appear near the edge of the polar pack ice.

In any scientific study, a factor paramount in the scientist's mind is accurate recording of the phenomenon involved. Anyone who has witnessed auroral displays would agree with the explorer Captain William Parry that it is impossible to record them adequately in words.

> Thus far description may give some faint idea of this brilliant and extraordinary phenomenon, because its figure here maintained some degree of regularity; but during the most splendid part of its continuance, it is, I believe, almost impossible to convey to the minds of others an adequate conception of the truth. It is with much difference, therefore, that I offer [a] description, the only recommendation of which perhaps is, that it was written immediately after witnessing this magnificent display. . . .

William E. Parry, *Journal of a Second Voyage for the Discovery of a North-West Passage from the Atlantic to the Pacific; Performed in the Years 1821-22-23, in His Majesty's Ships Fury and Hecla*, New York: Greenwood Press, 1904, p. 143

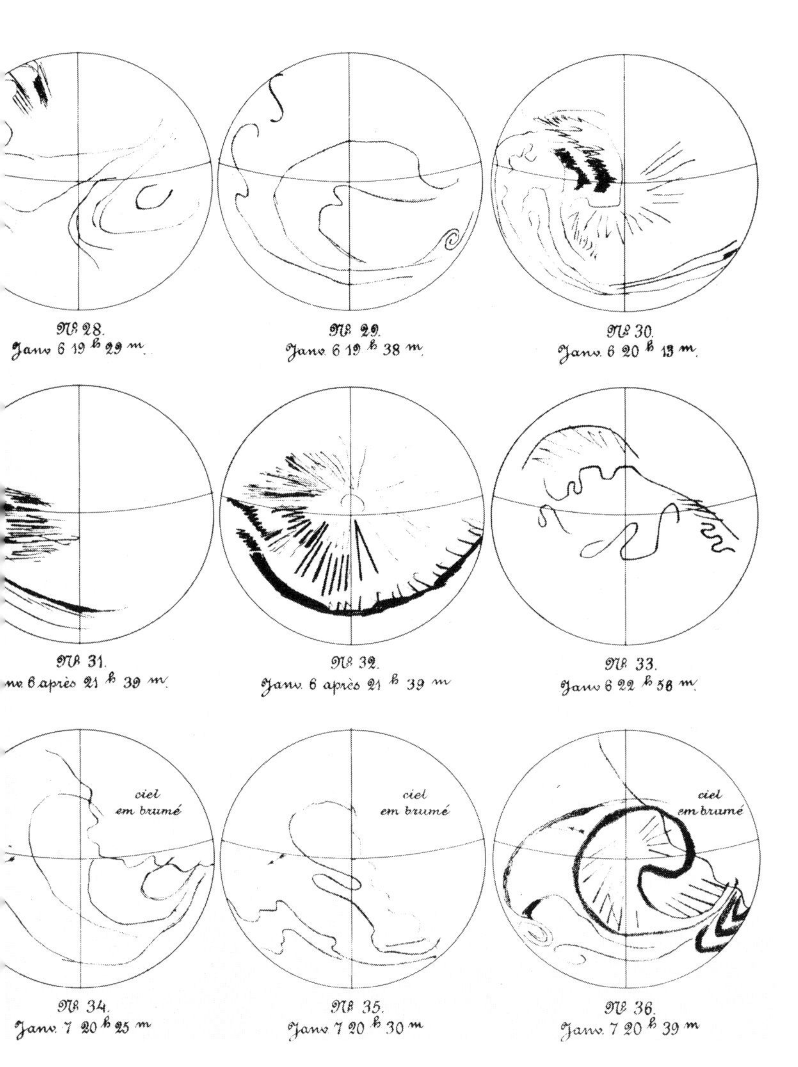

There are now a variety of recording systems employed in auroral studies, sophisticated electronic devices, often automated, as well as many photographic techniques. Obviously, such aids did not exist in early times and the scientists had to sketch the aurora, often with freezing hands. A drawing by De Mairan in 1726, is believed to be one of the first scientific sketches of the aurora. The whole sky is represented by a circle.

Photography was introduced into auroral study at the very end of the last century. The difficulties faced by scientists at that time are well expressed by Sophus Tromholt, an enthusiastic auroral observer:

> "Every attempt I made to photograph the Aurora Borealis failed . . . In spite of using the most sensitive dry plates, and exposing them from four to seven minutes, I did not succeed in obtaining even the very faint trace of a negative."

ABOVE — *Nine rough drawings of auroral formations. It must have been a difficult task to sketch rapidly changing auroral forms with freezing hands. (V. Carlheim-Gyllenskold)*

RIGHT — *Two pioneer auroral scientists, Kristian Birkeland and Carl Stormer, were photographing the aurora in the northernmost part of Norway in about 1905. (Courtesy of Auroral Observatory, University of Tromsø, Norway)*

Stormer made important contributions to auroral science between 1910 and 1930 on the basis of 40,000 photographs he took personally. They were used to determine auroral heights and forms. Veryl R. Fuller and Ervin H. Bramhall of the University of Alaska also made a large number of photographic observations of the aurora in Fairbanks around 1930-34.

There are two problems associated with auroral photography. The first is, as noted earlier, that auroral motions are often so violent that even our sophisticated modern films are inadequate to record them. A sensitive television system is required.

The second problem in recording the aurora is that the aurora is a very large-scale phenomenon, so that it is not possible to study it adequately at single points; observations from more than one point must be accurately coordinated. Since such coordination was not possible in the early days, one of the major unanswered questions during the first half of this century was how auroral curtains are distributed over the entire polar region, as would be seen when one looks down on the earth from a great distance above the North Pole. It was generally and tacitly believed that the auroral curtains lay along the auroral zone determined by Loomis.

During the 1957-8 International Geophysical Year

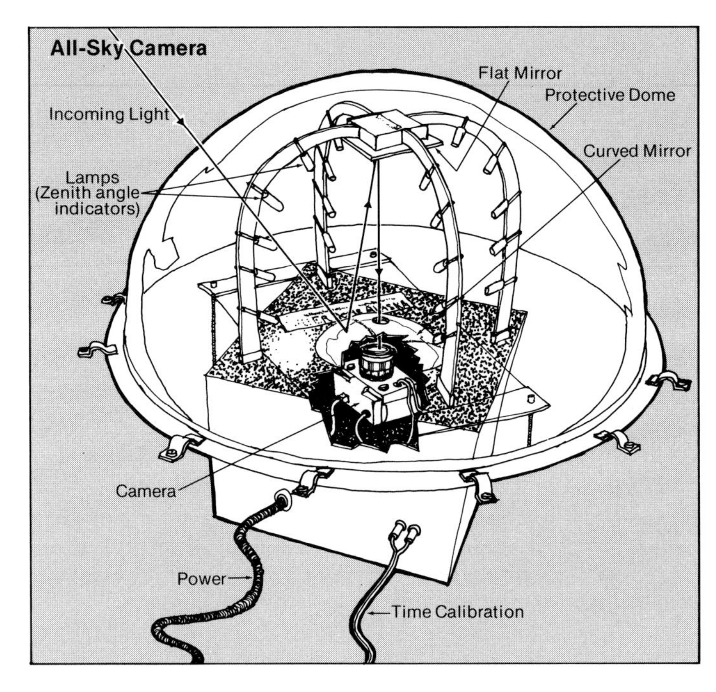

THE STRUCTURE AND OPTICS OF AN ALL-SKY CAMERA: *More than a hundred of these cameras were operated during the International Geophysical Year (1957-8). Based on all-sky photographs, auroral activity over the whole polar region was revealed. (Geophysical Institute, University of Alaska)*

(IGY), auroral scientists made an all-out effort to determine precisely the distribution. For this purpose, they devised a camera that is capable of photographing the entire sky in a single frame. Such a camera is called, appropriately, an "all-sky camera." At that time, a fast fish-eye lens was not available, so it was necessary to construct a camera system consisting of a convex mirror, which reflects any object at any point in the sky from horizon to horizon, and a flat mirror above the convex one, which reflects any image on the convex mirror. The camera is located in a box below the mirror system and takes a photograph of the image reflected on the flat mirror through a small hole at the top of the convex mirror. The convex mirror ''shrinks'' the zenith area, so the flat mirror above does not seriously block the view.

More than a hundred of these cameras were installed in the arctic and antarctic wilderness during the IGY in order to photograph auroras simultaneously one per minute. The resulting films were analyzed by a number of scientists, including Yasha I. Feldstein and O. V. Khorosheva of the University of Moscow. Feldstein and Khorosheva discovered in 1963 that auroras are distributed along a narrow band encircling the pole, called the auroral oval, which is quite different from the auroral zone. Suppose you could look down on the north polar region from above. You would see a beautiful annular belt of the aurora, the auroral oval, which is fixed with respect to the sun. The earth rotates under it once a day. Thus, the geographic pattern under the oval changes as the earth rotates, and unlike the auroral zone, the auroral oval is not fixed at a particular geographic location at all times. The auroral

A color photograph of active auroral bands taken by an all-sky camera. (Geophysical Institute, University of Alaska)

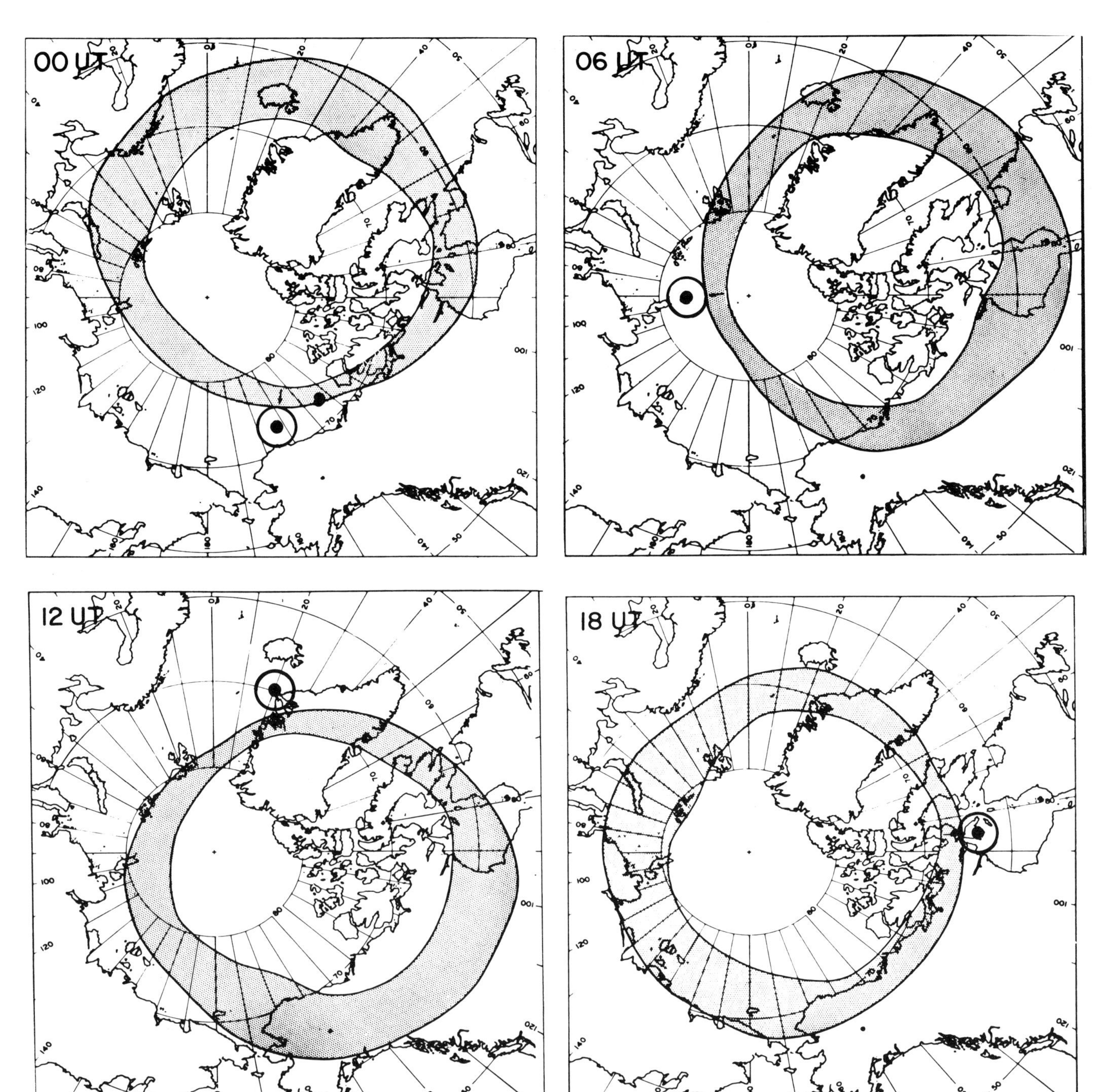

On a geographic map, the auroral oval is located at different places at different universal time (UT - Greenwich Mean Time). At 00 UT (2 P.M. in Interior Alaska), the midday part of the oval is located in the Beaufort Sea in the Alaska longitude; the direction of the sun is indicated by a dot with a circle. At 06 UT (8 P.M.), the oval is over Inuvik, Northwest Territories, Canada. At 12 UT (2 A.M.), the main part of Alaska is under the oval. At 18 UT (8 A.M.), the oval covers the North Slope region of Alaska. (Courtesy of Geophysical Institute, University of Alaska)

A special U.S. Air Force jet assigned to the Geophysical Laboratory at Hanscom Air Force Base, Bedford, Massachusetts, for auroral and polar ionospheric research. (Air Force Geophysics Laboratory)

zone is the locus of the midnight part of the auroral oval on the earth as it rotates once a day.

This concept of the auroral oval was quite controversial for several years after it was first suggested and was not necessarily accepted by a number of auroral scientists who clung to the belief that the aurora was distributed along the auroral zone. Since the distribution of the aurora was an important issue, not only to basic science, but also in a practical way due to its effect on radio communications, two jet aircraft were assigned to the problem. One, a U.S. Air Force jet, was stationed at the Geophysical Laboratory, Hanscom Field, Bedford, Massachusetts; the other aircraft was the NASA jet *Galileo*. The presence of the auroral oval was confirmed by an intensive analysis of photographs taken from the two planes.

S.-I. Akasofu (right) and Al McNeil in the instrumented cabin of the NASA jet Galileo. *(Ames Research Center, NASA)*

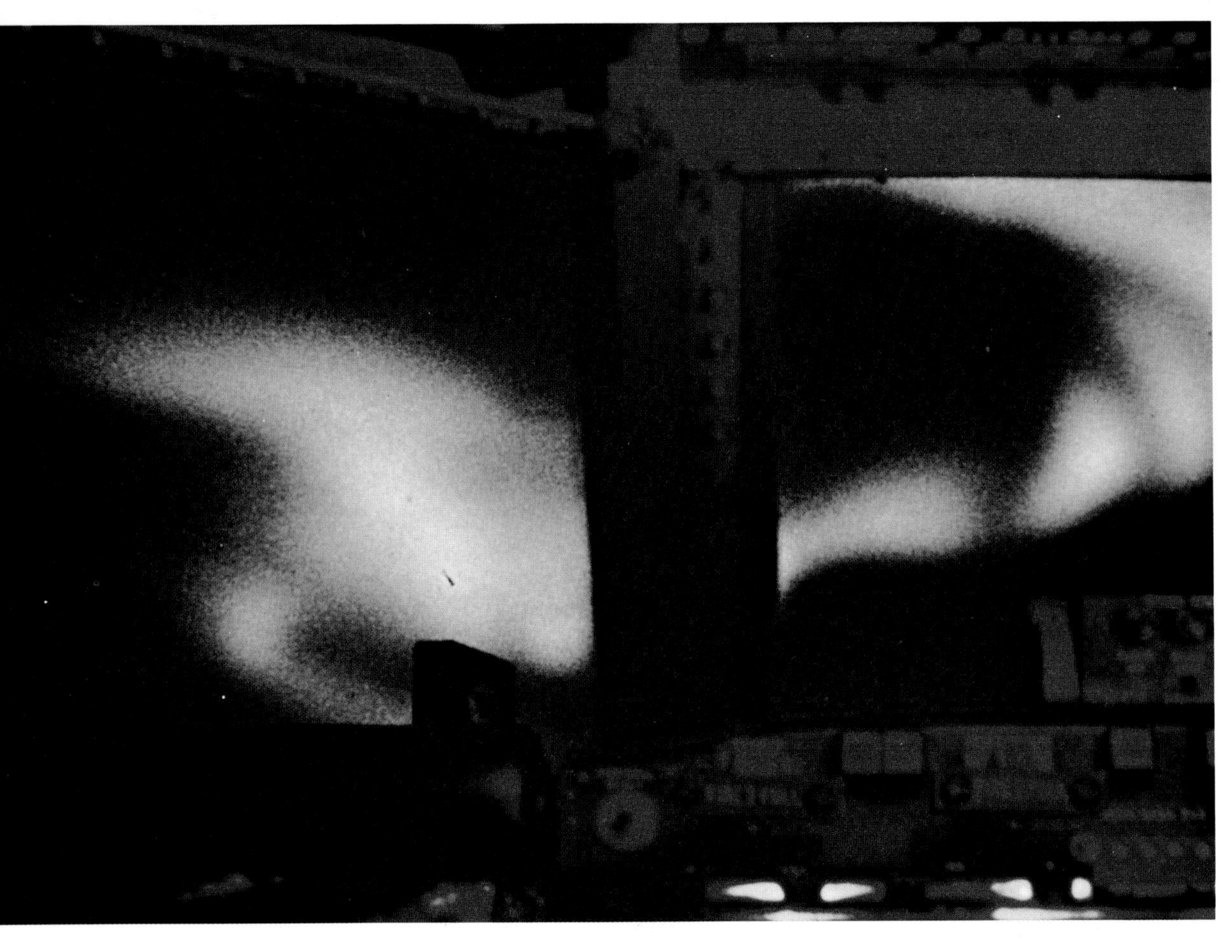

ABOVE — *An auroral display photographed from the cockpit of* Galileo. *(Ames Research Center, NASA)*
RIGHT — *The aurora borealis lights the sky over the wing of* Galileo. *(Geophysical Institute, University of Alaska)*

Two heavily instrumented jets also investigated the question of whether the auroras in the northern and southern hemispheres are the same or in some way different. In 1967 scientists from the Geophysical Institute, University of Alaska and from the Los Alamos Scientific Laboratory, one group flying over Alaska and the other flying simultaneously well south of New Zealand, proved that the aurora borealis and the aurora australis are essentially the same. Photographs show the two similar

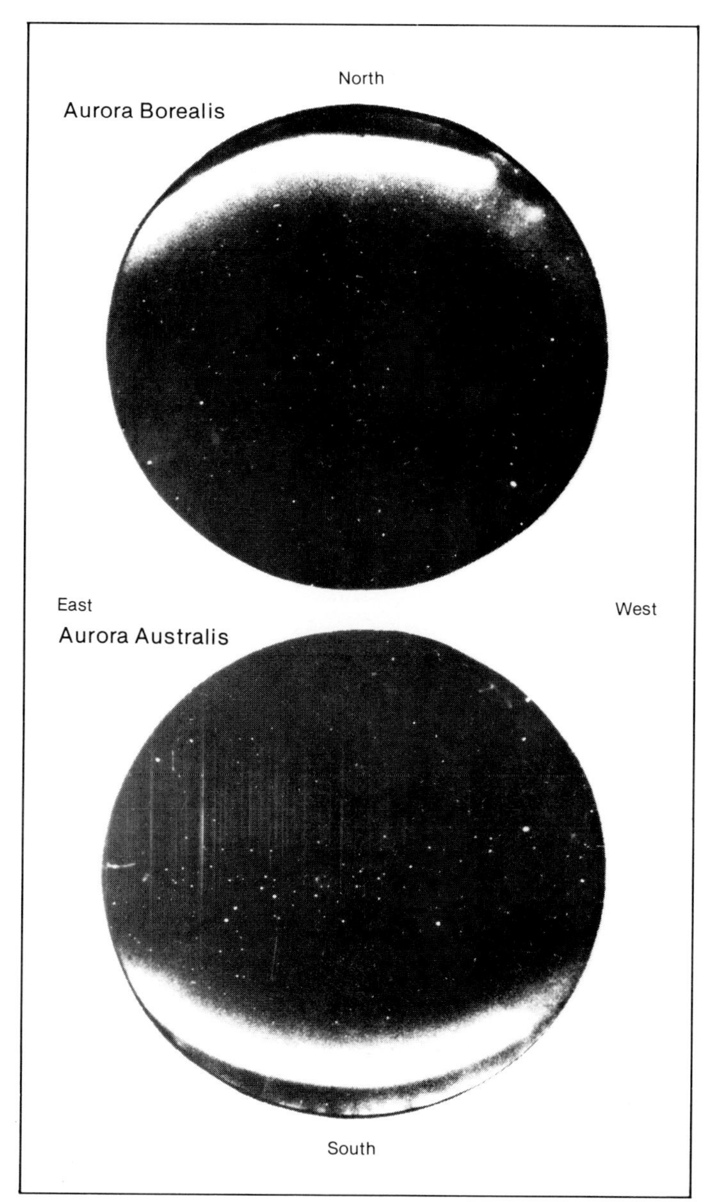

LEFT — *The aurora in the Northern Hemisphere (aurora borealis) and the aurora in the Southern Hemisphere (aurora australis) are essentially the same. These two all-sky photographs were taken simultaneously by two jet aircraft, one flying over Alaska (upper) and one well south of New Zealand (lower). (A.E. Belon, J.E. Maggs, T.N. Davis, K.B. Mather, N.W. Glass and G.F. Hughes,* Journal of Geophysical Research, *1969, courtesy of S.-I. Akasofu)*

ABOVE — *Auroral display over the Japanese Antarctic Station, Syowa. One can again see that the aurora australis has basically the same features as those of the aurora borealis. (T. Oguti, Geophysical Institute, Tokyo University)*

auroral bands. A single magnetic field line connected the centers of the northern and southern areas shown in the photographs (see the next chapter).

With the 1970's came the "space age" for auroral photography. A Canadian satellite, ISIS-II, was the first to carry a device for photographing the aurora from high above the polar region. Now, several satellites are conducting such a photographic study.

Another important question is, What kind of light does the aurora emit? The answer can tell us two important things. The first is the kinds of atoms and molecules that are emitting lights, and the second is why those atoms and molecules emit lights. This particular field of science is called spectroscopy, or in our specific case, auroral spectroscopy. The simplest instrument for spectroscopy is a prism; the analyzed pattern of light produced through a prism is called a spectrum.

Until early in the 19th century, it was commonly believed that the aurora was caused by the reflection of sunlight from tiny ice crystals in the sky. If that were so,

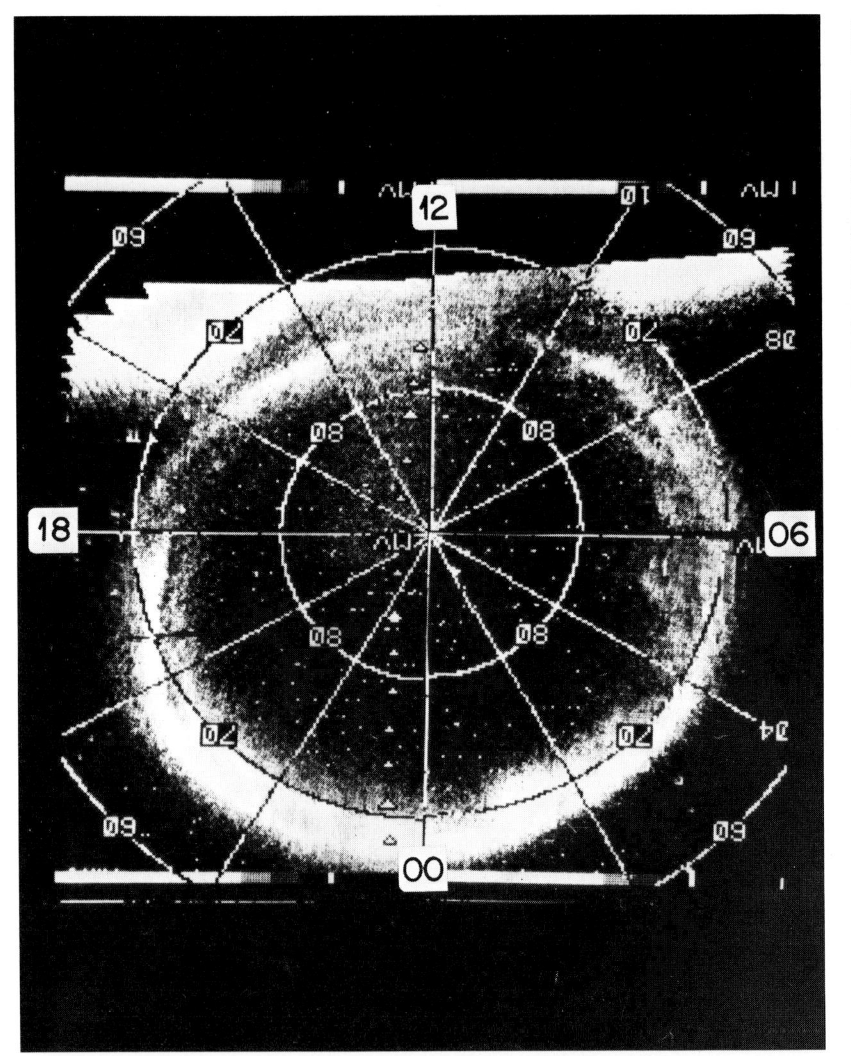

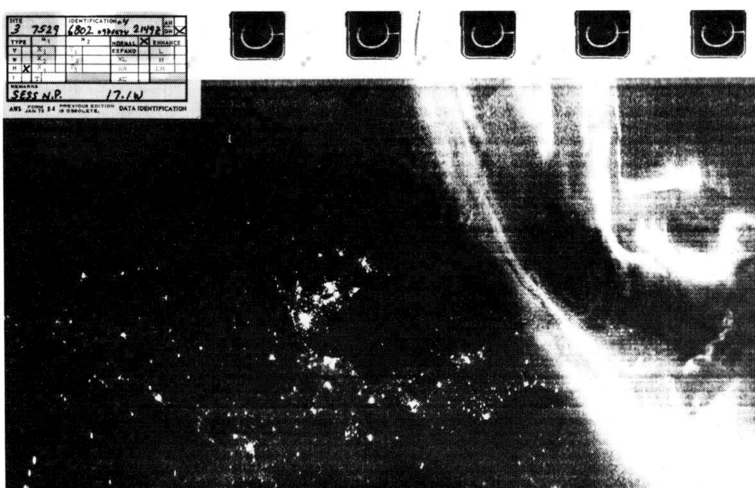

LEFT — *One of the first photographs of the auroral oval taken from above the north polar region by a satellite. (Cliff Anger, University of Calgary)*
ABOVE — *Satellite photograph of the auroral oval over northern Europe by the USAF satellite (Defense Meteorological Satellite Program). The largest bright spot is the city of London. (World Data Center-A, NOAA)*

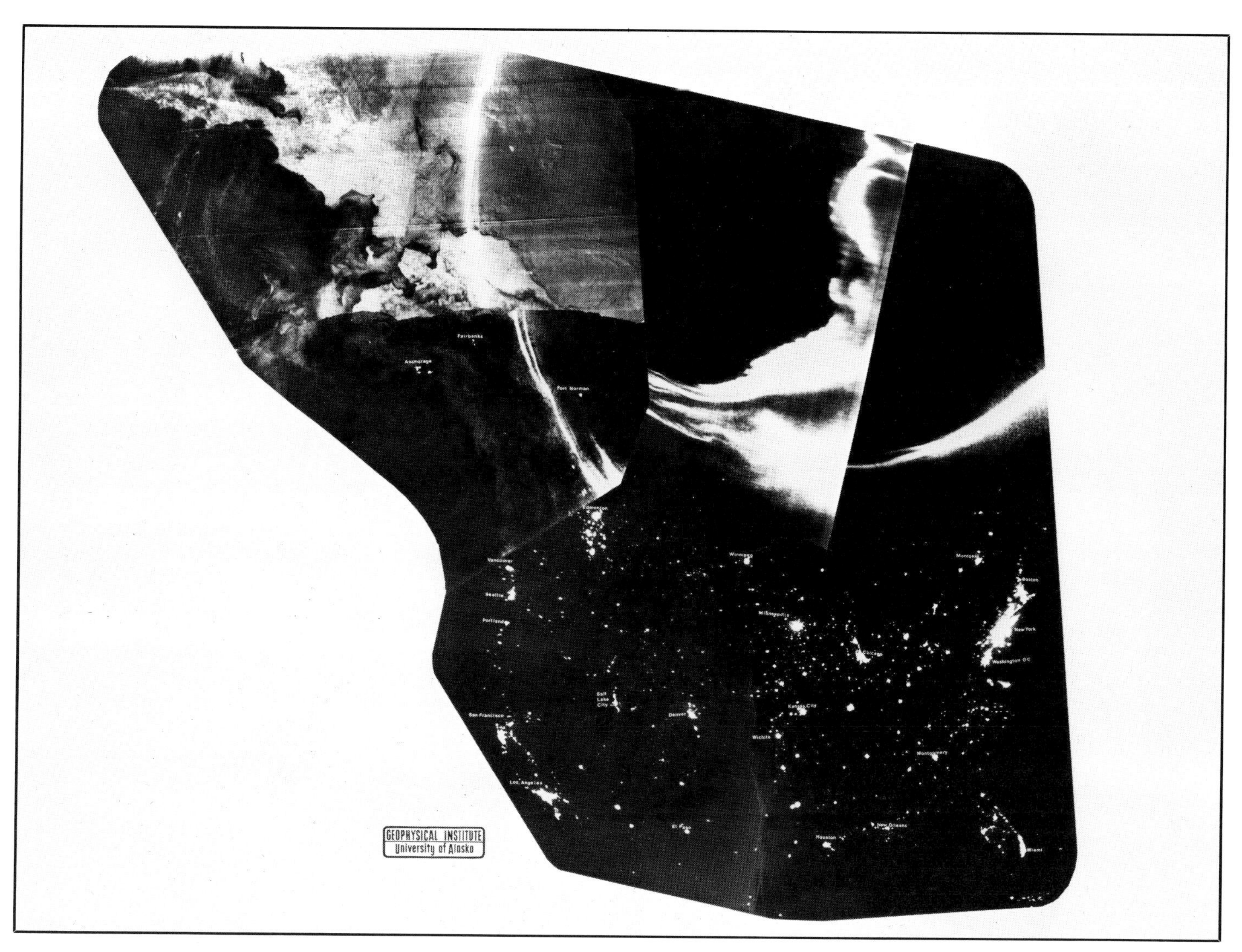

Montage photograph of North America taken by the USAF satellite (Defense Meteorological Satellite Program). Since several photographs taken at different times are combined, the auroral activity appears to be discontinuous. (World Data Center-A, NOAA)

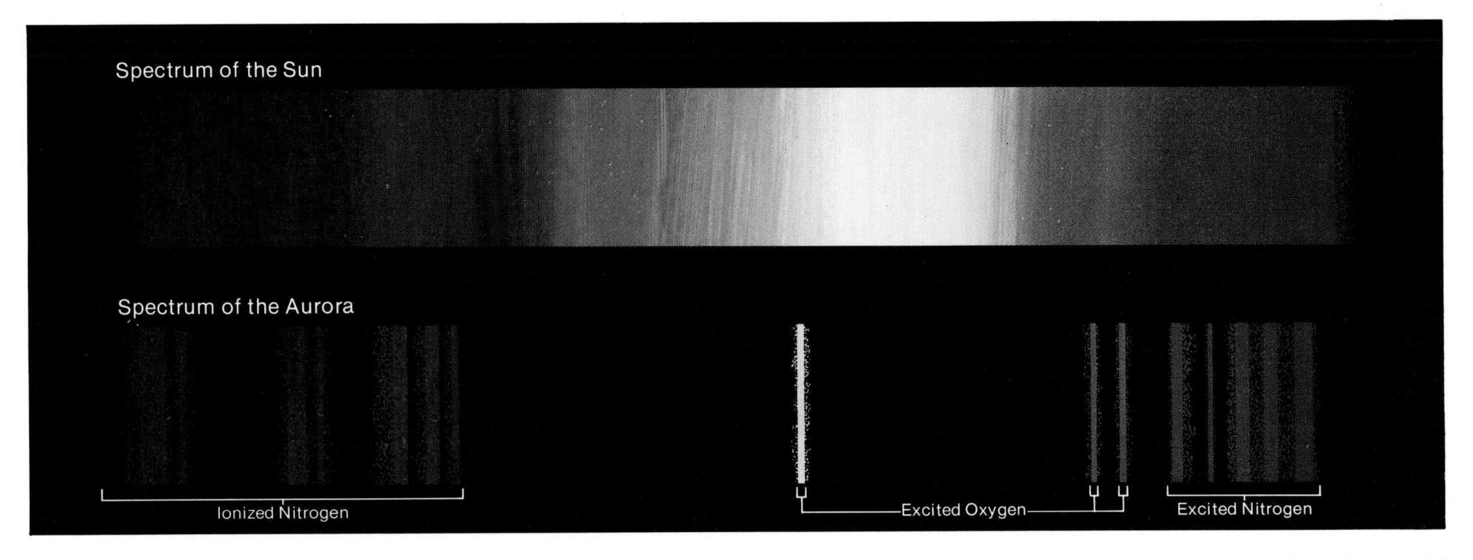

COMPARISON OF THE SPECTRUM OF SUNLIGHT PRODUCED BY A PRISM AND A TYPICAL SPECTRUM OF AURORAL LIGHT: *The solar spectrum shows the familiar rainbow colors, a continuous transition from red to violet, while the auroral spectrum consists of many lines and bands of different colors.*

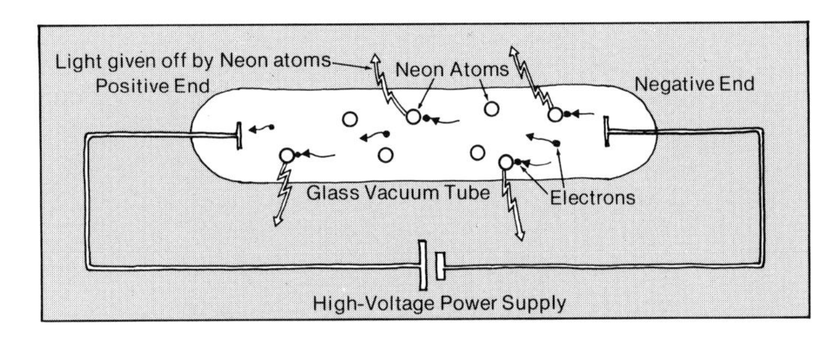

THE PRINCIPLE OF A NEON SIGN: *A bright red light is emitted from neon atoms in a vacuum glass tube with electrodes at both ends when a high voltage is applied between them—an electrical discharge phenomenon.*

when the auroral light is seen through a prism, it should show the familiar rainbow colors, which range continuously from red to violet; such a spectrum is thus called a continuous spectrum.

Norwegian physicist Anders Jonas Angstrom (1814-1874) was one of the first to use a prism to study the aurora; he found that auroral light is nothing like that of a rainbow. The auroral spectrum is not continuous and it consists of many lines and bands of different colors with dark spaces among them. The lines are emitted by atoms and the bands by molecules.

In the middle of the 19th century, physicists were aware that a spectrum consisting of such lines and bands of light can be produced from a vacuum glass vessel when a high voltage is applied through electrodes inserted into the vessel. These lights come from the atoms and molecules remaining in the vessel when they are struck by high-speed electrons which are emitted from the negative electrode connected to the high-voltage source. The electrons carry the current from the negative electrode to the positive one. This process is in common use in the form of the neon sign.

To illustrate this effect, a thin glass tube in which a vacuum has been induced is filled with a small amount of neon gas, and is then connected to a high voltage source. Electrons stream from the negative end to the positive end of the glass tube. The neon atoms in the way are hit by the streaming electrons and their internal state changes. This process is called excitation. However, a neon atom cannot remain excited for very long and must return back to the "ground state," or normal condition. It is during this process of return to ground state that the familiar bright red light is produced. This particular emission is a characteristic color of neon atoms, and no other kinds of atoms can emit the same light. Similarly, when mercury vapor is sealed in the tube and excited, a bright green light is produced. Again, no other kinds of atoms can emit this same color light. Physicists have compiled a complete list (or "spectrum") of the lights emitted by different kinds of atoms and molecules, organized by wavelength. Wavelength has been divided into "Angstrom" units, with one Angstrom defined as equal to 0.00000001 cm. In this way we can now recognize most of the atoms and molecules which emit auroral lights in the polar upper atmosphere.

LEFT — *The auroral electrical discharge takes place in the upper atmosphere. (M. Lockwood, Geophysical Institute, University of Alaska)*
BELOW — *When electrical discharge tubes filled with neon gas and mercury vapor are connected to a high-voltage source, each emits its own characteristic light. (Geophysical Institute, University of Alaska)*

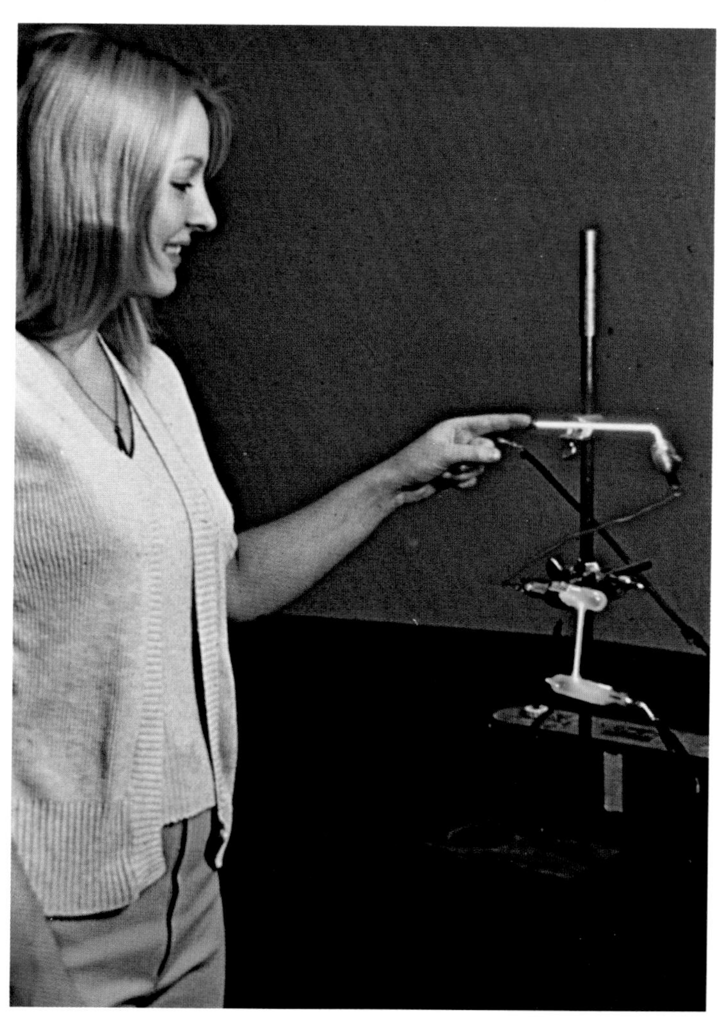

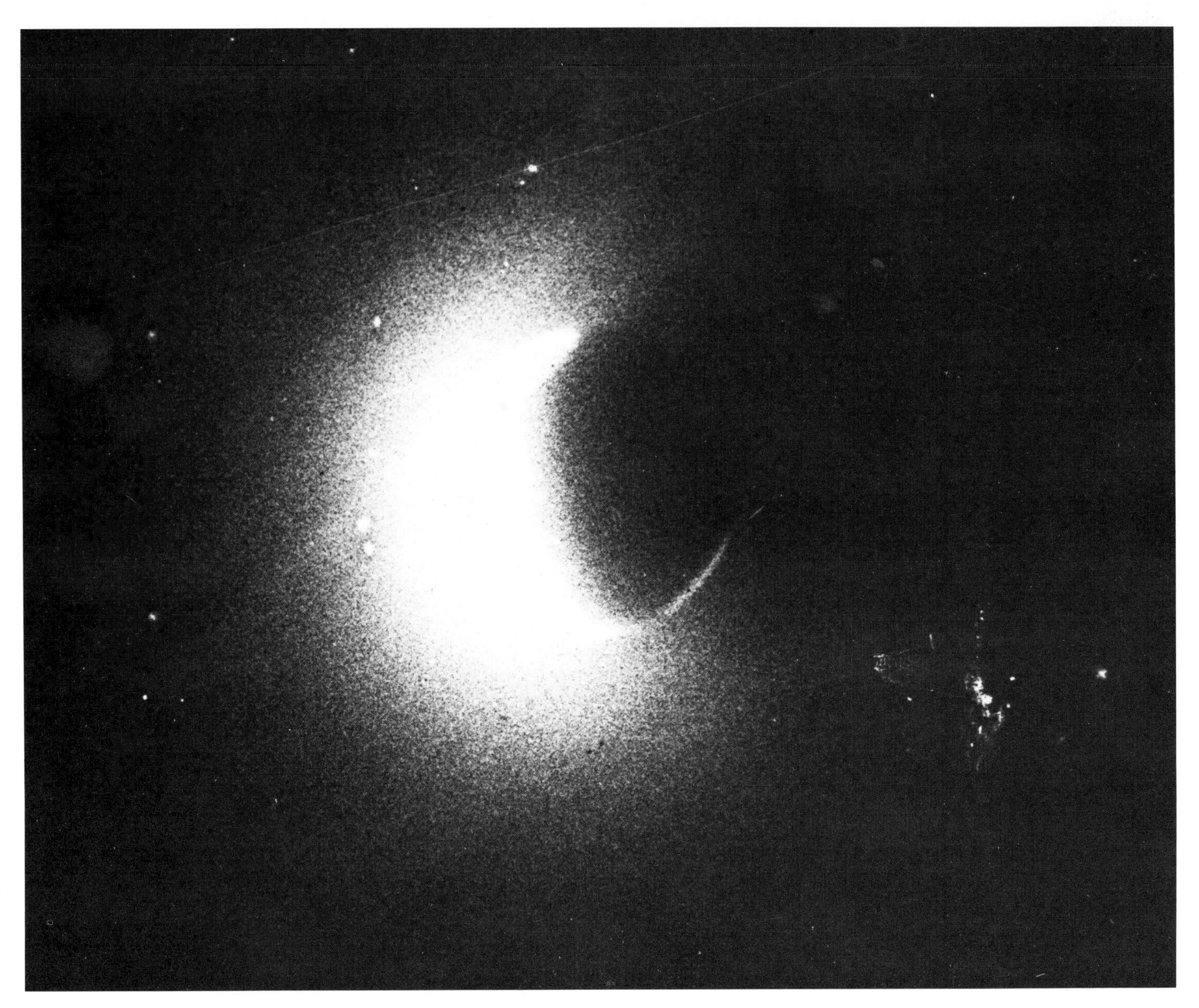

An ultraviolet photograph of earth taken from the moon by astronauts during the Apollo 16 mission. Both the aurora borealis (top) and the aurora australis (bottom) can be recognized. (G.R. Carruthers and T. Page)

Angstrom found that the most common light form of the aurora, a whitish green color, had a wavelength of 5567 Angstrom (the present accurate determination is 5577 Angstrom). This line is called the green line. From the time of Angstrom's discovery (1868), to as late as 1924, this green line was a mystery because no one could successfully find an atom which produced it. In 1925, two Canadian scientists finally discovered that this whitish green light is emitted from atomic oxygen. In the lower atmosphere, oxygen exists in a molecular form (O_2). However, at the height at which the aurora appears, the oxygen molecules are found separated into their two constituent oxygen atoms (O). Atomic oxygen can also emit a dark red light (6300 Angstrom) under some circumstances—that ''bloody red color'' which caused so much fear in medieval days. This line is called the red line.

ABOVE — *Lars Vegard, one of the pioneers of auroral spectroscopy, at work in northern Norway. His instrument is called the auroral spectrograph. (Courtesy of University of Oslo)*
LEFT — *Anders Jonas Angstrom, who found that the auroral spectrum is very different from the solar spectrum. (Courtesy of W. Stoffregen)*

When the aurora becomes very active, the bottom of the folded curtain is often tinted a crimson red, one of the most beautiful colors of the aurora. This light is a band and is emitted by molecules. Scientists identified nitrogen molecules, consisting of two nitrogen atoms (N_2), as being responsible.

Lars Vegard, a Norwegian scientist, was the pioneer in the field of auroral spectroscopy. During the first half of this century, auroral light was studied in great detail by scientists from the United States, Canada, Scandinavia, the U.S.S.R. and elsewhere.

Auroral spectroscopy tells us that the aurora is a "dis-

The "bloody red" light is emitted by atomic oxygen. During a great magnetic storm, this color is considerably enhanced. This dramatic photograph makes it easy to understand why the appearance of this type of aurora caused so much fear in medieval days. (Geophysical Institute, University of Alaska)

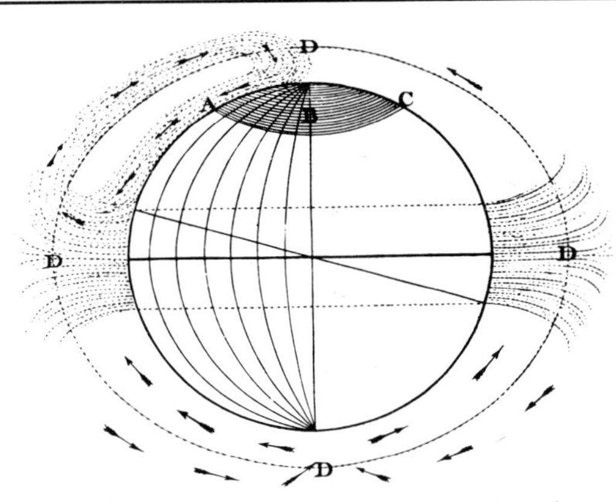

The Arrows represent the general Currents of the Air.
ABC. the great Cake of Ice & Snow in the Polar Regions.
DDDD. the Medium Height of the Atmosphere.
The Representation is made only for one Quarter and one Meridian of the Globe; but is to be understood the same for all the rest.

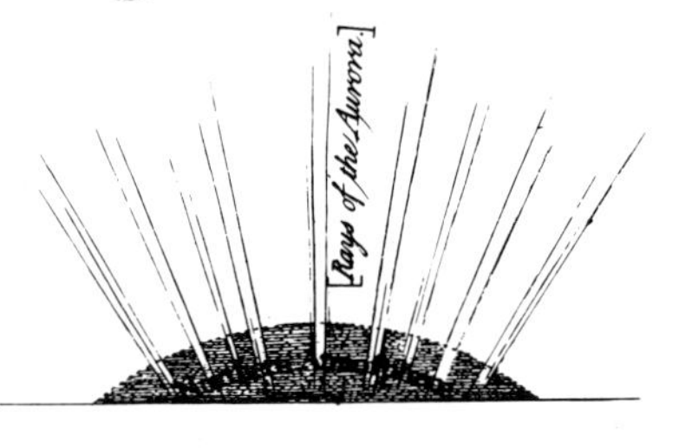

Benjamin Franklin's diagram showing his conception of the circulation of the earth's atmosphere and its relation to the aurora. (Courtesy of Franklin Institute)

If the primary energetic electrons are energetic enough, they can penetrate down to an altitude of about 90 km (56 mi.). However, much of their energy is lost by the time they get down to such a low altitude, so that they are incapable of ionizing nitrogen molecules. But they can still excite the nitrogen molecules and cause them to emit the crimson red color.

Many people have reported hearing a crackling or hissing sound coming from the aurora on occasion. Clarence A. Chant, for instance, reported in 1923, "We watched this display approaching from the north. At first there was no sound, but as it got nearer, we heard a subdued swishing sound, which grew more distinct as it approached, and was loudest when the ribbon or belt of light was right overhead." Another observer, D. M. Garber, wrote in 1933, "The spectacle was so awe inspiring that a dog team was stopped, and I sat upon the sled for more than an hour absorbing the marvelous beauty of this most unusual display. As we sat upon the sled and the great beams passed directly over our heads, they emitted a distinct audible sound which resembled the crackling of steam escaping from a small jet." (S. M. Silverman and T. F. Tuan. *Auroral Audibility,* Advance in Geophysics, Vol. 16, edited by H. E. Landsberg and J. Van Meighen. New York: Academic Press, 1973, pp. 156-266.)

In order to study such a "phenomenon," auroral scientists must, first of all, record and analyze it. Unfortunately, scientists have not been able to record any sound even with modern, sophisticated audio equipment. As early as 1885 Tromholt concluded, "Without absolutely refusing to believe in the possible existence of such a sound, I fancy that there must be some acoustic deception or misunderstanding which has created this belief in an auroral sound." Auroral scientists have discovered that an intense auroral display does produce very low frequency pressure pulses (infrasonic), but their frequency is too low to be detected audibly.

The aurora emits both ultraviolet and infrared lights and x-rays. Both the ultraviolet and x-rays are absorbed in the atmosphere and thus do not reach the ground. The aurora

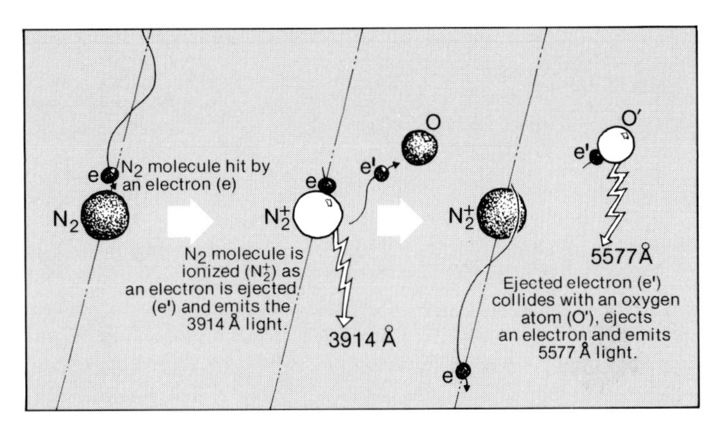

TRIGGERING THE AURORAL LIGHT: *An energetic electron (e) enters the upper atmosphere and collides with a nitrogen molecule (N_2). The electron passes through the molecule and in the process "kicks out" one of the molecule's electrons (e^1), ionizing the molecule. The ionized nitrogen molecule (N_2) then emits a strong violet light. The secondary electron (e^1) is also energentic and may in turn strike an oxygen atom (O). The oxygen atom becomes "excited" and emits the greenish white light we see as the aurora.*

also emits radio waves over a very wide frequency range, called radio "noise" which can be heard by a radio receiver, but not by humans. It is quite intense in the broadcast band (500-1600 kHz). Fortunately, the ionosphere is shielding us from this radio noise. Without the ionosphere, we would not be able to use a radio, and from outer space the earth would be heard as a very "noisy" planet.

"Fred's Electric Nursery," powered by the aurora. (From the Voyage of the Vivian, *T. W. Knox, 1884, courtesy of S.-I. Akasofu)*

Solving the Riddle of the Aurora

It has now become clear that the auroral lights appear when atoms and molecules in the upper atmosphere are hit by high speed electrons. Indeed, we shall see in this chapter that the aurora is a gigantic electrical-discharge phenomenon surrounding the earth. How is this electrical discharge powered? What kind of processes in the sky can act like a generator, supplying the electricity to this most spectacular natural phenomenon? The electrical power associated with the auroral discharge is enormous, about 1,000 billion watts, or annual 9,000 billion kilowatt hours — more than the present annual U.S. electric power consumption, which is a little less than 1,000 billion kilowatt hours!

Even as early as the 19th century some people were quite certain that the aurora was an electrical phenomenon and thought about the possible use of auroral electricity.

> "When we have time to spare we will set about devising a machine whereby the electricity of the aurora borealis may be harnessed, and made to do duty in a practical way. We will make it run the dynamos to supply our houses and streets with electric light; it shall propel our machinery, and thus take the place of steam; it shall be used for forcing our gardens, in the way that electricity is supposed to make plants grow; and it shall develop the brains of our statesmen and legislators, to make them wiser and better and of more practical use than they are at present. Hens shall lay more eggs, cows must give cream in place of milk, trees shall bear fruit of gold or silver, tear-drops shall be diamonds, and the rocks of the fields shall become alabaster or amber. Wonderful things will be done when we get the electricity of the aurora under our control."
>
> "Yes," responded Fred, "babies shall be taken from the nursery and reared on electricity, which will be far more nutritious than their ordinary food. When the world is filled with giants nourished from the aurora, the ordinary mortal will tremble. We'll think it over, and see what we can do."
>
> Thomas W. Knox, *The Voyage of the Vivian*, 1884

Early ideas about the cause of the auroral discharge were primitive. Not surprisingly, Benjamin Franklin proposed an electrical discharge theory of the aurora. He thought

Edmund Halley's diagram of the earth's magnetic field, 1716. (Courtesy of S.-I. Akasofu)

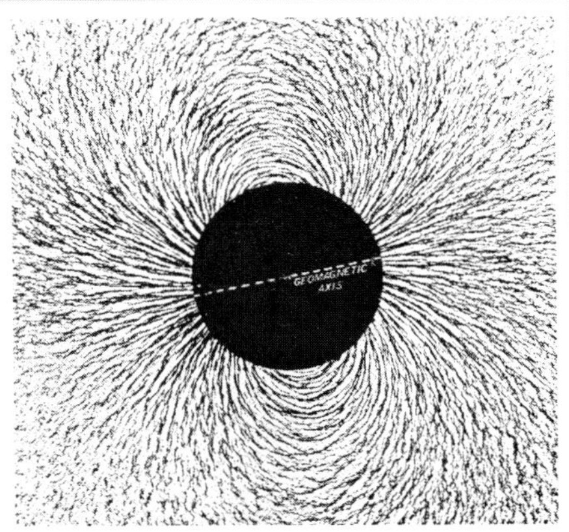

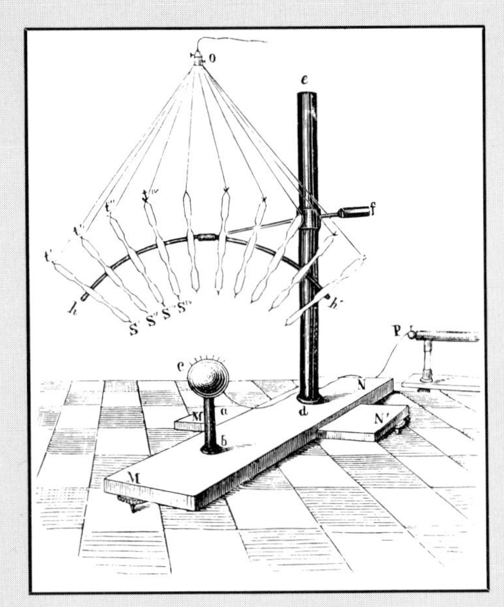

ABOVE — *Magnetic field lines. When a magnet is placed beneath a sheet of paper on which iron filings have been sprinkled, the filings tend to line up along imaginary lines, called magnetic field lines (From* Geomagnetism *by Chapman and Bartels, 1940)*

UPPER RIGHT — *Lemstrom's apparatus with which he attempted to simulate the aurora in 1879. (From* L'Aurora Boreale *by S. Lemstrom, Paris, 1886)*

RIGHT — *The Norwegian physicist Kristian Birkeland (left) and his assistant Olav Devik in his laboratory with a big vacuum box in which he attempted to reproduce the auroral zone. (Courtesy of University of Oslo)*

The first diagram of the earth's magnetic field was constructed by William Gilbert (1544-1603), physician to Queen Elizabeth I and one of the most distinguished men of science during his era. In a book published in 1600, he declared, "The earth itself is a gigantic magnet." A little more than a century later, in 1716, Edmund Halley, best known to us as the discoverer of the great comet that carries his name, produced the second diagram of the earth's magnetic field in its ideal form. In Halley's diagram, solid and dotted lines represent imaginary magnetic lines of force or magnetic field lines. This can be shown by the familiar demonstration of placing a magnet on a sprinkling of iron filings. The filings will tend to align themselves along these imaginary field lines.

It is not difficult to suppose that the magnetic field of the earth is a part of the needed ingredient in the auroral discharge phenomenon.

What, then, can play the role of the conductor? It turns out that this component is the solar wind, high-speed charged particles consisting mostly of protons and electrons in equal number, streaming out from the sun. More precisely, the conductor is the outermost part of the sun's atmosphere, called the corona, which appears during a total solar eclipse as a beautiful light around the masked sun. (Don't look at the sun with the naked eye at any time, except at the time of the totality; serious eye damage can result.) The corona has a very high temperature (1 million degrees C) so that all the atoms and molecules that make it up are ionized (that is, some of the electrons in the atoms are stripped off) and thus they are charged particles. We know that a stream of charged particles is a conductor. Further, the solar wind contains magnetic fields which originate from the sun. (Sunspots, too, have strong magnetic fields.) Imagine magnetic field lines in the solar wind being stretched out or carried away as the solar wind expands outward from the sun. Such a gas of charged particles, consisting of equal numbers of positive particles and negative particles (mostly electrons), is called a plasma. It is often referred to as "the fourth state of matter," the first three being solid, liquid and gaseous (neutral).

When the solar wind "blows" away from the sun or reaches the vicinity of the earth, a cavity, shaped like a comet, is formed around the earth. This is because the earth's magnetic field becomes an invisible obstacle to the solar wind; thus the solar wind flows around the front of the obstacle. The cavity is called the magnetosphere. The distance from the earth to the nose of the cavity is about 65,000 kilometers (40,365 miles). This would be all we would expect if the solar wind did not "carry out" the magnetic field from the sun; there would not be any aurora on the earth.

The magnetic field of the solar wind plays a very important role. This is something auroral scientists began to understand only 10 years ago. What happens is that some of the magnetic field lines in the solar wind get connected to a bundle of the magnetic field lines of the earth. One can imagine a bundle of magnetic field lines above the polar region (called the polar cap), which fans out in the cavity and gets connected to the magnetic field lines of the solar wind across the boundary of the magnetosphere.

Now, as mentioned earlier, the solar wind blows around the boundary of the magnetosphere so that it has to blow across the connected field lines. It is in this way that electricity is generated, because a conductor (the solar wind) moves across a magnetic field. Actually, the entire boundary surface of the magnetosphere is the generator, called the solar wind-magnetosphere generator. There is a man-made generator, called the MHD generator, which works on the same principle. Auroral scientists have found that the power generated by this action is about 1,000 billion watts, with a voltage of about 100,000 volts.

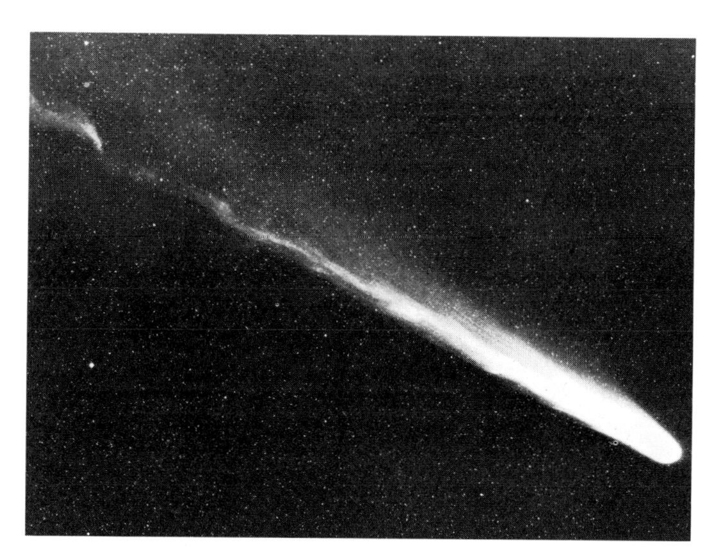

Comet Kohoutek, like most comets, exhibits a similar kind of shape as the cavity called the magnetosphere. This cavity is formed around the earth when the solar wind "blows" in our vicinity.
(Courtesy of Hale Observatories)

A generator must have two terminals. For our solar wind-magnetosphere generator, the positive terminal is located on the morning side of the boundary surface of the magnetosphere and the negative terminal on the evening side. In order to bring the power produced by the generator to the polar upper atmosphere to have a discharge, the upper atmosphere must be connected to the terminals by "wires." In a rarefied ionized gas such as that which fills the magnetosphere, electric currents can flow more easily along magnetic field lines than perpendicular to them, so that the magnetic field lines become "invisible wires."

The bundles of the earth's magnetic field lines from the polar region (called the polar cap) are connected to the

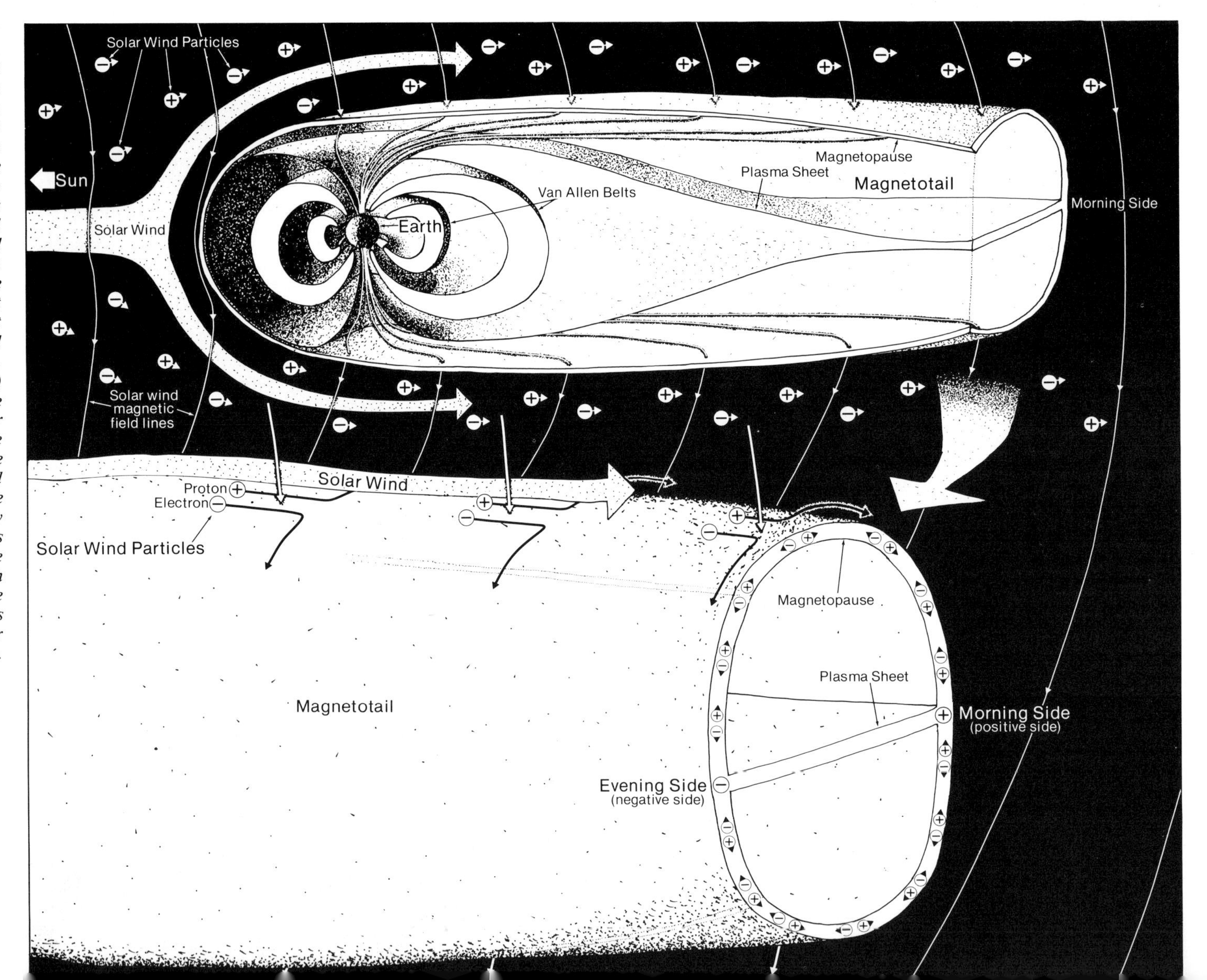

THE MAGNETOSPHERE AS AN ELECTRIC GENERATOR: *The magnetosphere looks like a cylindrical shell with a blunt nose. In it, there is an interesting structure, such as the Van Allen belt and the plasma sheet. The Van Allen belt has a doughnutlike structure and its nightside becomes flattened, forming a sheet called the plasma sheet. The solar wind particles blow around the magnetosphere, forming a comet-shaped cavity. The earth is located near the head of the cavity. The cylindrical cavity extends anti-sunward at least a length of 1000 earth radii. This portion is called the magnetotail. In this illustration, we have cut open the magnetotail to take a close look at its boundary (called the magnetopause) and the motions of solar wind particles there. Solar wind particles consist of protons (+) and electrons (-). As these blow along the magnetopause, they are deflected by the solar magnetic field. The protons tend to move around the magnetopause toward the morning side in both the Northern and Southern hemispheres. As a result, the morning side of the plasma sheet is positively charged. On the other hand, the electrons tend to move around the magnetopause toward the evening side, again in both hemispheres. As a result, the side of the plasma sheet is charged negatively. This is the basic feature of the solar wind-magnetosphere generator.*

solar wind magnetic field lines. Among these field lines, only the field lines making up the surface of the bundle are connected to the terminals, so that the electric current from the terminals of the generator flows from the positive terminal to the polar upper atmosphere and back to the negative terminal only along the surface of the bundle of the polar cap. The current is mainly carried by electrons. When these electrons collide with upper atmospheric atoms and molecules, the atoms and molecules emit their own characteristic lights, which we recognize as the aurora (diagram page 66).

Imagine a funnel-shaped electron beam surrounding the bundle of magnetic field lines from the polar cap, so that the earthward end of this funnel-shaped region glows, as

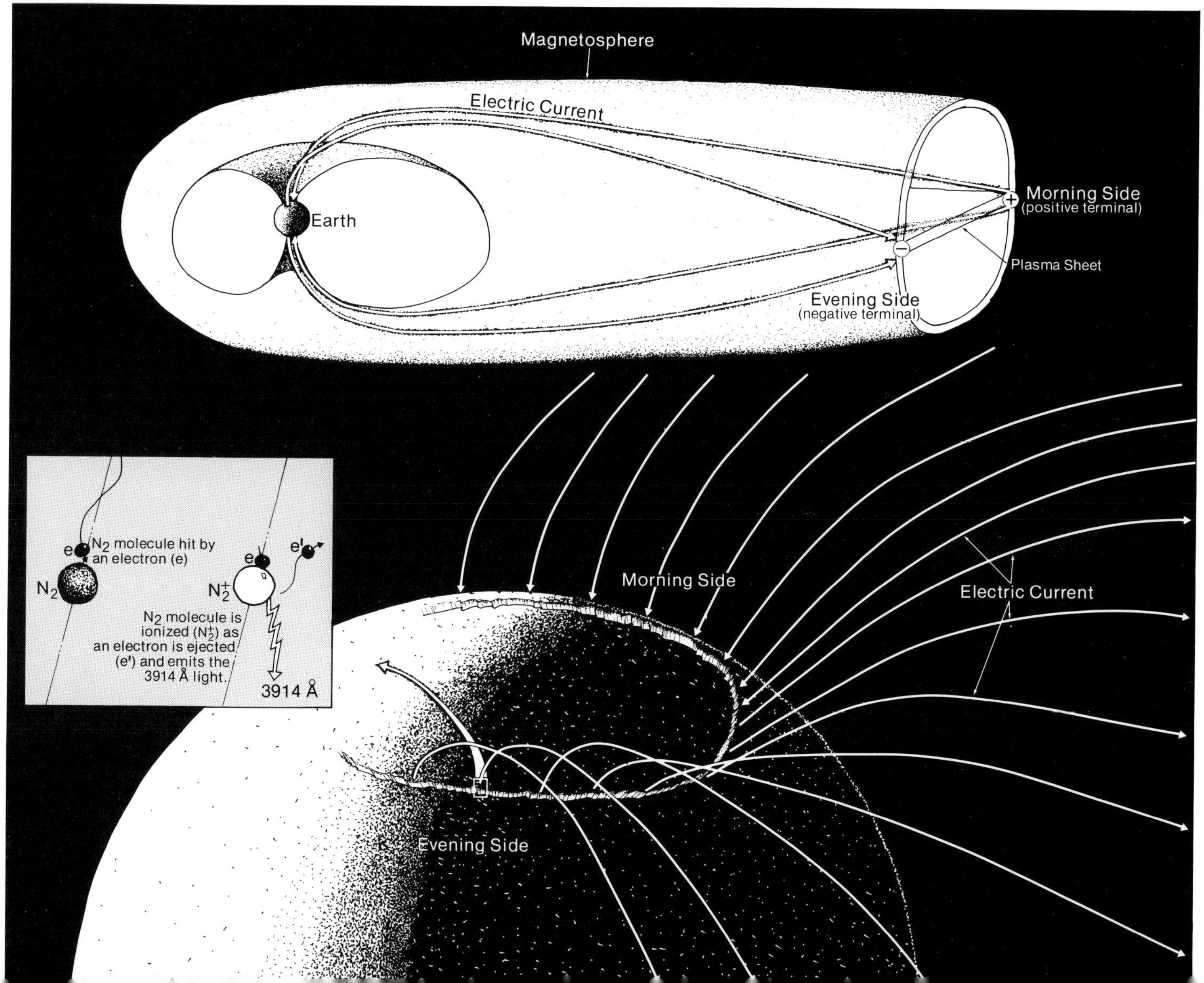

THE AURORAL ELECTRIC CIRCUIT: *The morning side of the plasma sheet becomes the positive "terminal" of the solar wind-magnetosphere generator and the evening side of the plasma sheet becomes the negative "terminal." The electric current flows from the positive terminal to the earth (the polar upper atmosphere) in the morning side and back from the earth to the negative terminal in the evening side. Looking closely at the earth, you can see the auroral curtain where the electric current from the solar wind-magnetosphere generator enters into the polar upper atmosphere and leaves from it. In the evening side, the upward current from the ionosphere is carried by downcoming electrons which ionize or excite upper atmospheric particles. The auroral lights are emitted by such electrons.*

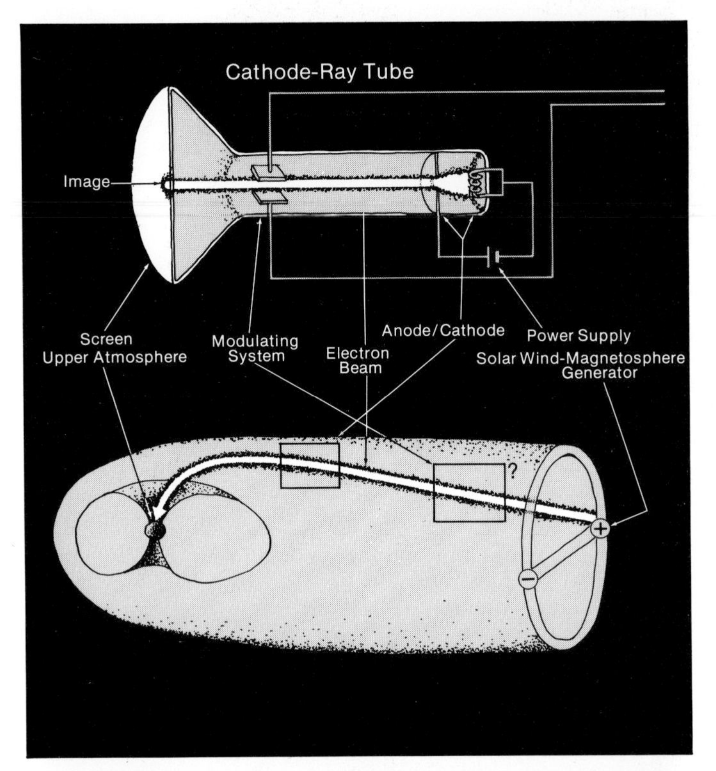

THE SIMILARITY OF THE PRINCIPLE BETWEEN A CATHODE-RAY TUBE AND AURORAL PHENOMENA: *Auroral scientists are trying to uncover various physical processes which correspond to various components of a cathode-ray tube.*

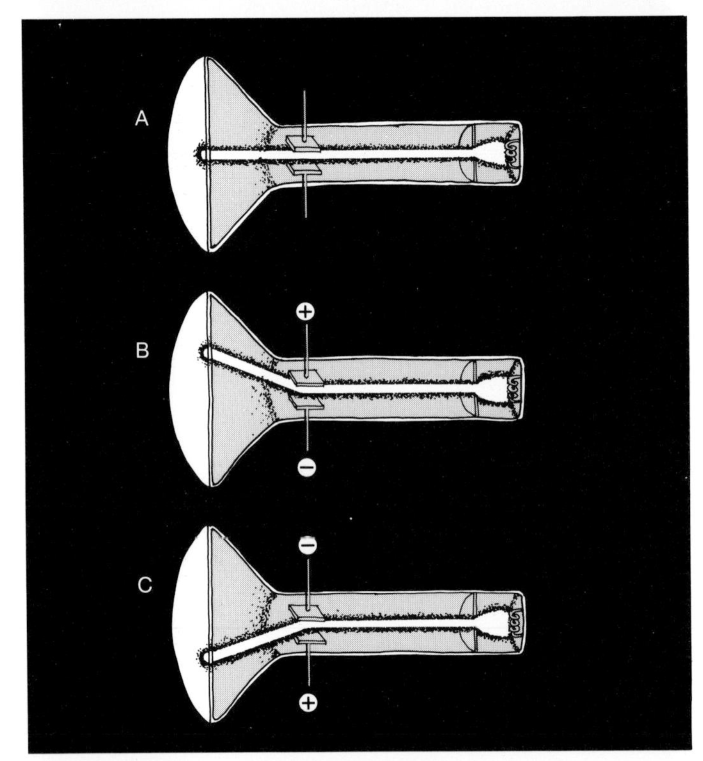

HOW THE IMAGE ON THE SCREEN OF A CATHODE-RAY TUBE MOVES: *Auroral motions are caused by complex changes of electric and magnetic fields in the magnetosphere, which are now under intensive study by auroral scientists on the basis of data taken from satellites and ground-based observatories.*

altitude. Electrons of energies 3-10 kilovolts are stopped at altitudes between 100 and 110 kilometers (62 and 68 miles).

The aurora is not just a glowing spot in the sky. It appears along the auroral oval, the bright ringlike belt surrounding the polar region which has already been described. In a cathode-ray tube are various devices which can modulate the electron beam on its way to the screen in order to form a particular shape of image. Both electric and magnetic fields are used to shape the electron beam to produce the image desired. In an equivalent manner, both the earth's magnetic field and the electric field (generated by the solar wind-magnetosphere generator) surrounding the earth modulate the electron beam to produce the auroral oval and the curtainlike or ribbonlike forms of the aurora.

The aurora is a very dynamic phenomenon. From time to time it shows a violent display. This corresponds to the movement of the image on the screen of the cathode-ray tube. In the tube, the simplest motion of an image can be produced by changing the intensity of the electric field between two plates between which the electron beam must pass on its way to the screen. When voltage is applied with the positive pole at the top plate and the negative pole at the bottom as shown in (B) on the lower left figure, the electron beam, which is always negatively charged, will be bent upward because opposite poles attract. As a result, the impact point of the electron beam on the screen shifts upward, and we see this as an upward motion of the image. Such a change of polarity of the electric field can be achieved by applying an alternating voltage to the plates (like our household power supply drawn from the power line). Similarly, when we shift the polarity and apply the opposite voltage to the plates (C) the electron beam will be bent in the opposite direction. Now we see a downward motion of the image. If no voltage is applied to the plates, the image will not be bent at all (A). A coil which produces a magnetic field can be used to deflect the electron beam from side to side in the same way.

Through the same principle, auroral motions are caused by changing electric and magnetic fields in the magneto-

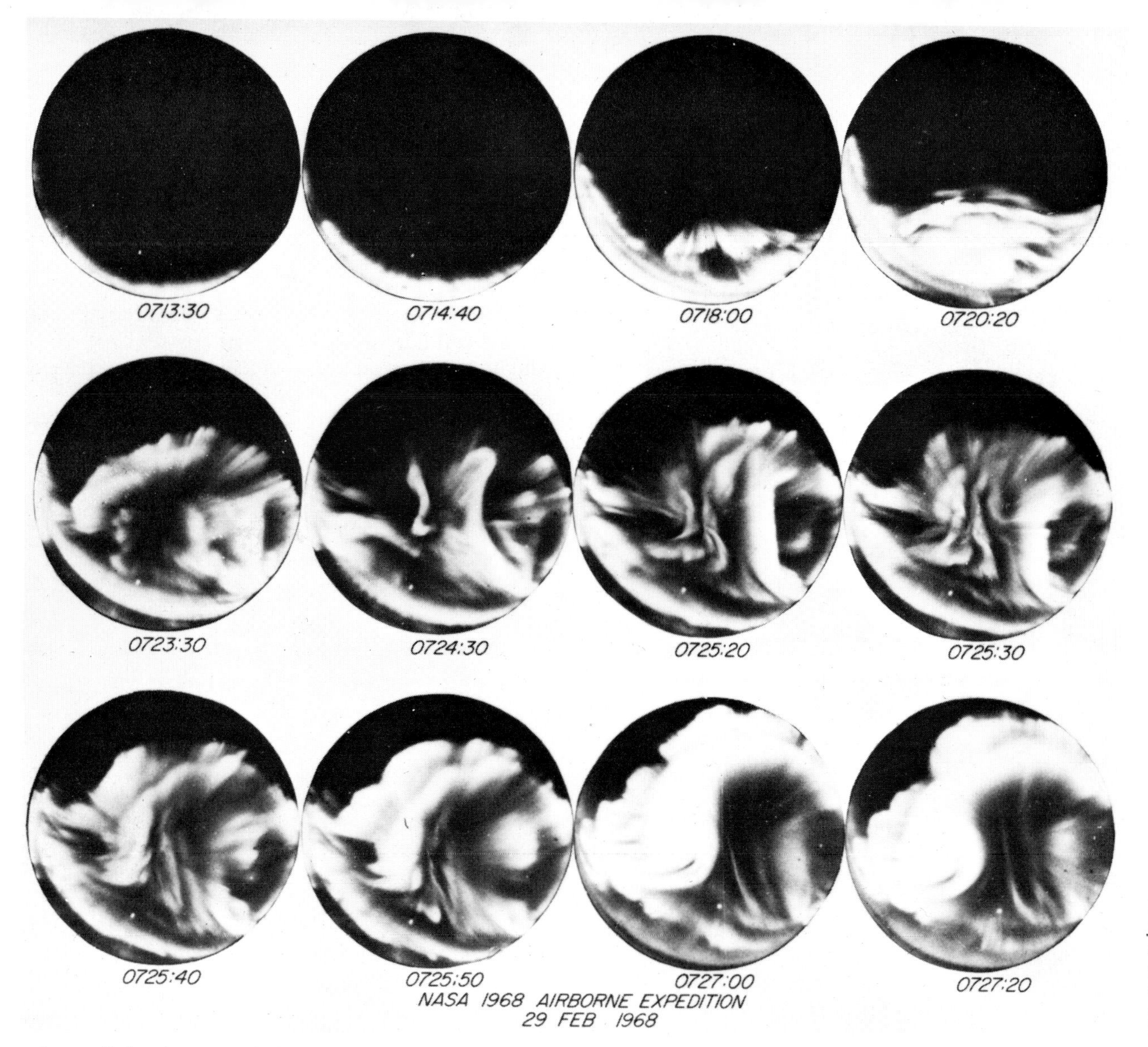

Timed all-sky photographs taken from the NASA jet Galileo. *This series of photographs shows how rapidly auroral activity spreads and changes over the sky. (Geophysical Institute, University of Alaska)*

sphere. Scientists can infer certain characteristics of the changing magnetic and electric fields on the basis of a study of auroral motions.

As the early polar explorers recounted, auroral displays are infinitely variable and complex. However, an extensive analysis of photographs taken during the International Geophysical Year by a series of all-sky cameras showed that auroral motions also have interesting regularities. By further inference, then, we can assume that the electric and magnetic fields in the magnetosphere must undergo some *systematic* variations. Now, through analysis of the data obtained from both satellites and ground-based observatories, scientists have begun to identify the changing patterns of these electromagnetic fields.

In the most common auroral activity, the initial action is seen as a sudden brightening of an auroral curtain in the midnight sky at the observer's meridian. As its brightness increases, the ray and wavy structures predominate. In a matter of a few minutes, a northward motion begins. This northward motion produces a large-scale wave structure, called the westward traveling surge, which moves rapidly toward the west along the auroral curtain. Often such a surge culminates in a spectacular curling motion of the auroral curtain accompanied by a crimson tint. Meanwhile, the midnight sky is filled with "broken curtains," a phenomenon also often referred to as the "break-up." Finally, in the morning sky, the auroral curtain disintegrates into patchy luminosity. Eventually, all auroral activity ceases, but a new chain of activity will eventually be repeated in the same way.

Such a systematic, quasi-cyclic auroral activity is called an auroral substorm. It lasts typically for about two or three hours. Details of such large-scale auroral motions are studied with the help of photographs taken from satellites. The eight photographs on the right illustrate how the development of an auroral substorm might be observed from a satellite. Imagine that you are high above the earth, watching the area indicated by a rectangle in the insert. The photographs show how the aurora would change in approximately half an hour.

We must now answer the question of why electric and magnetic fields in the vicinity of the earth change. (We are not talking here about the earth's own magnetic field.) The ultimate source of the changes is the sun. Changes in conditions on the sun cause changes in the solar wind and space. Solar wind changes in turn are reflected in changes in the efficiency of the solar wind-magnetosphere generator. Auroral scientists know that the speed and the magnetic field of the solar wind control the efficiency of the generator. The generator efficiency is proportional to the solar wind speed, and is also proportional to the square of the intensity of the solar wind magnetic field. Further, the direction of the solar wind magnetic field is vitally important. If the solar wind magnetic field is pointing northward, the generator's efficiency is at its minimum,

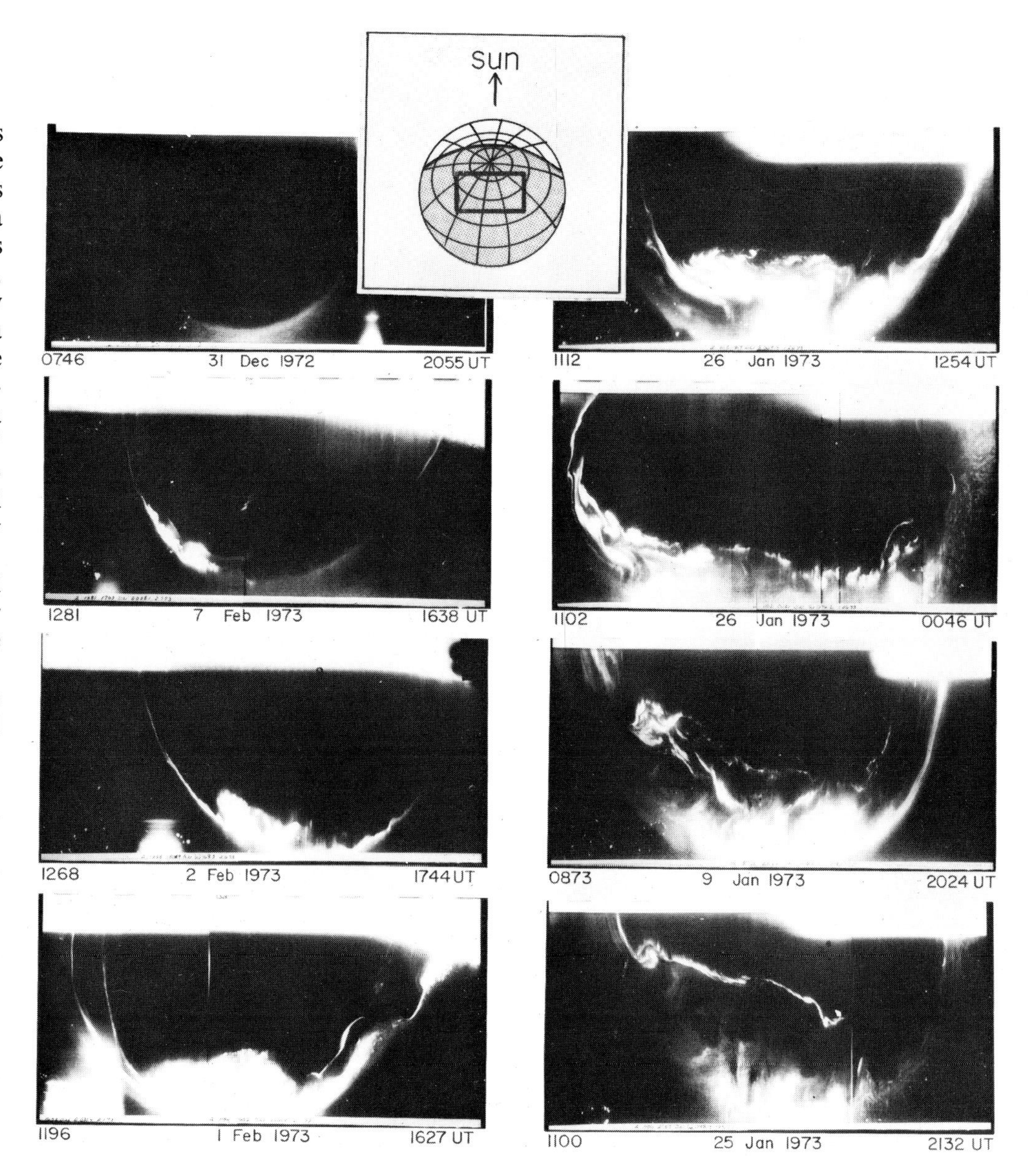

The development of an auroral substorm. If you could watch auroral displays from above the north polar region aboard a satellite, you would witness the development of auroral activity as this figure illustrates.
(The USAF Defense Meteorological Satellite Program)

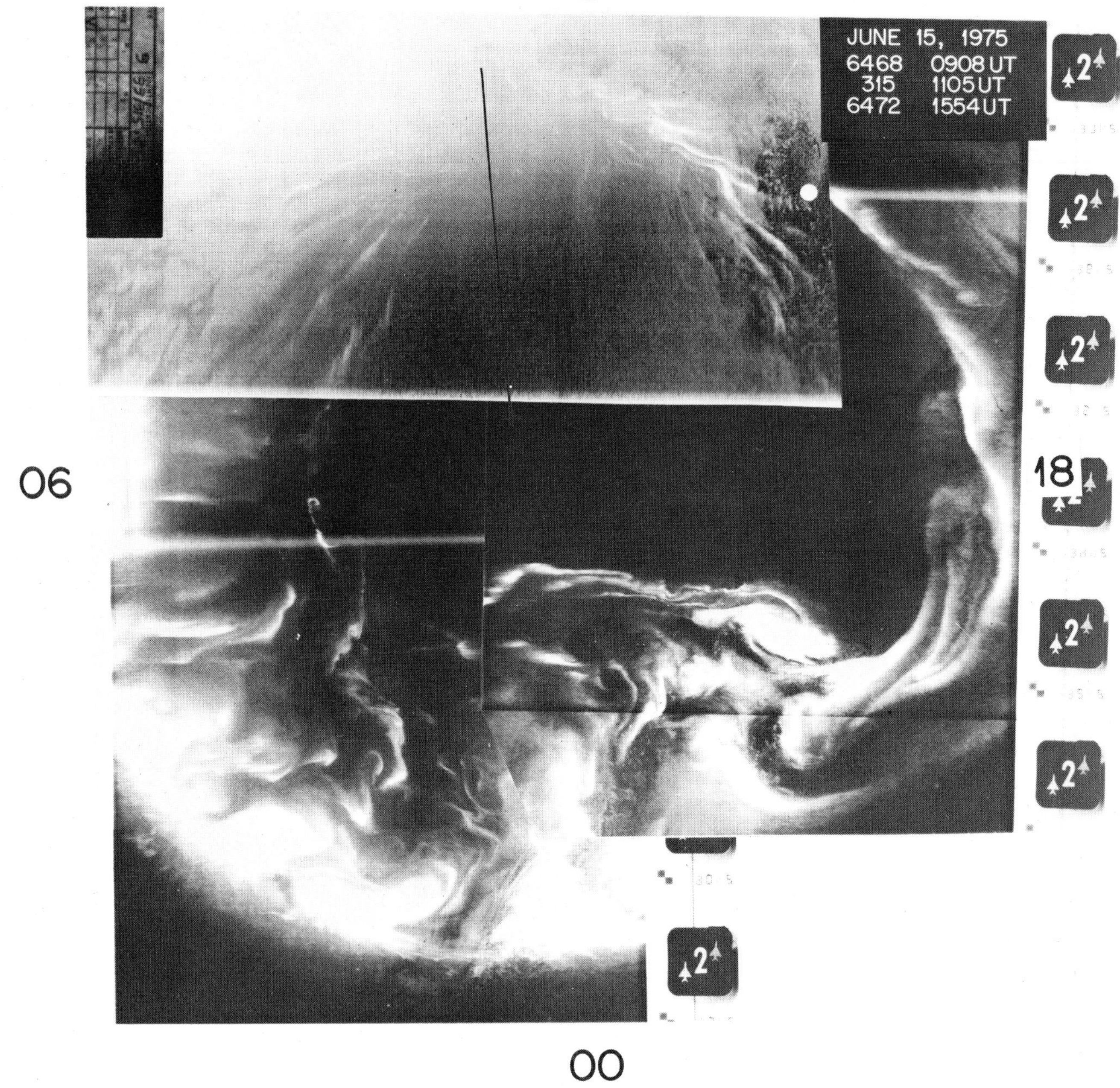

A montage photograph of the auroral oval over the Antarctic. Since this is a photograph of the Southern Hemisphere, the morning side is seen on the left, the evening side on the right. The top part is the noon sector, and the bottom part is the midnight sector. Note in particular an intense auroral activity in the night sector. (USAF Defense Meteorological Satellite Program)

and if it is pointing southward, the efficiency grows to maximum. When the generator efficiency is higher, the generated voltage and currents are higher, resulting in brighter auroras.

This process, therefore, not only explains what causes the changes in electric and magnetic fields in the vicinity of the earth, but also how auroral displays are related to conditions on the sun. The solution to the problem of the relationship of the aurora to the sun, however, did not come easily. It is based on efforts made by many scientists over the last hundred years. Even as recently as the beginning of this century it was not at all obvious that the aurora is related to the sun.

The story of the search for the earth-sun relationship began at 11:20 A.M. on September 1, 1859, when Richard C. Carrington, an English astronomer and solar physicist, was sketching sunspots. To his surprise, a very bright spot appeared among the spots, and he made the first notation of what we now call a solar flare, an intense explosion on the sun. Carrington "hastily ran to call someone to witness the exhibition." About one day later, a large part of Europe was covered by active auroras. In fact, there is also some record to indicate that the aurora was seen as far south as Honolulu on that day. Carrington suggested hesitantly that there was a possible connection between the solar explosion he had witnessed and the great auroral display, but he also warned against jumping to conclusions by remarking, "One swallow does not make a summer."

Many physicists and astronomers gave more heed to his warning than his suggested connection. Even in 1895, William Thomson, Lord Kelvin, one of the most famous British physicists of his day, commented that Carrington's finding was simply coincidental and called the problem a "fifty years' outstanding difficulty."

In 1905, after an exhaustive study of geomagnetic disturbances, another British solar physicist, Edward W. Maunder stated, "The origin of our magnetic disturbances lies in the sun." However, his colleague, the mathematician and physicist Sir Arthur Schuster, immediately criticized him, saying, "The mystery is left more mysterious than ever."

Many more years had to pass before the majority of scientists became convinced of the solar-terrestrial relationship, despite the fact that it had been well established in the latter part of the 19th century that the occurrence of the aurora has a similar cycle to that of the sunspot cycle. In 1884 Sophus Tromholt quipped, "How can this connection between what may be called a terrestrial phenomenon and solar disturbances be explained? Well, that it is one of those riddles which the scientist of today leaves to that of the future to solve." It turns out that the answer to this riddle is quite complicated, but a solution may now be coming into sight.

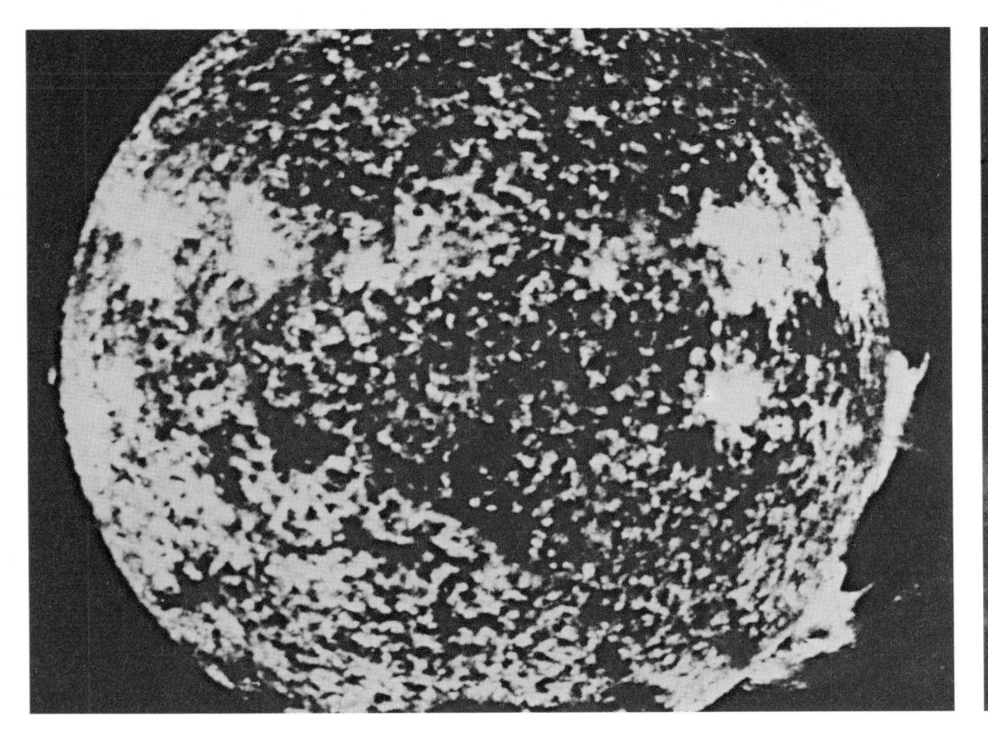

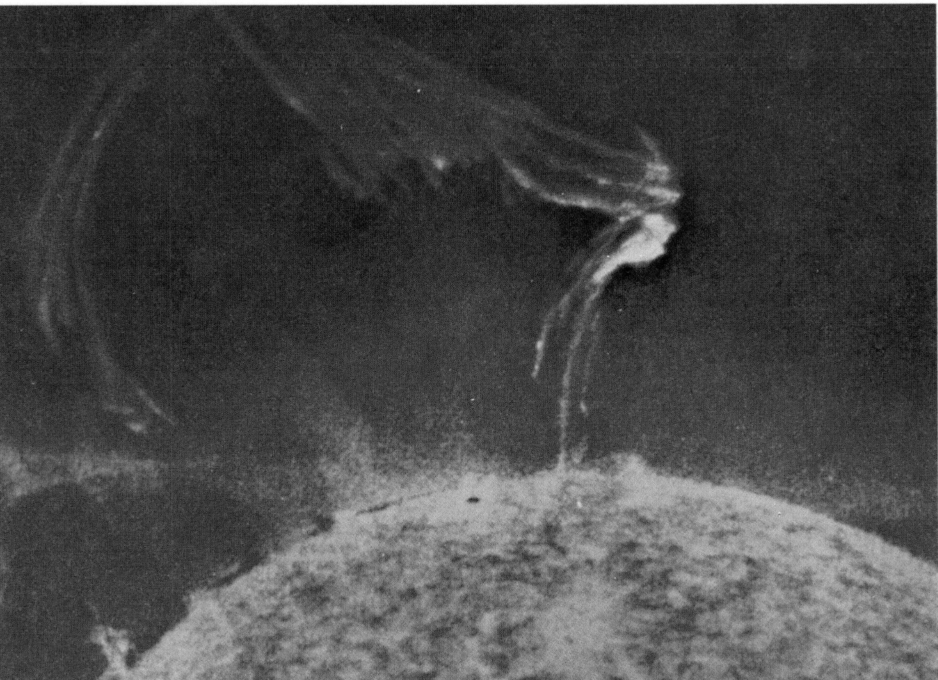

Gigantic prominences rising from the sun, photographed from Skylab. (Both photos by NRL and NASA)

Today we know that the solar wind is often considerably intensified after a solar flare. When an intense explosion takes place in an active region in the vicinity of a large sunspot group, a large amount of gas is ejected. This exploding gas may be termed the gusty solar wind. If the earth happens to be in the line of motion of this gas, it will take it 25 to 48 hours to reach the earth. This will occur when a flare is seen near the center of the solar disk; if a flare occurs near the edge of the disk, the gusty solar wind blows in a direction 90 degrees away from the earth, so that we expect little effect from such a flare.

Solar physicists have also found that a high-speed solar wind can originate from a relatively large area that is free from sunspots. This area is most clearly seen in soft x-ray photographs taken from Skylab and appears as a dark region. For this reason, the area is called a coronal hole. A coronal hole often persists for a few months to as long as a year. Since the sun rotates approximately every 27 days, a coronal hole faces the earth once in 27 days. And for this reason auroral and geomagnetic activity tends to recur every 27 days.

Thus, the correlation between the occurrence of the aurora and sunspots is not simple, although great auroral displays in mid-latitudes are usually associated with intense flares which occur in the vicinity of large sunspots.

When the earth is immersed in either the gusty solar wind or the high-speed wind, the efficiency of the solar wind-magnetosphere generator is increased, and bright auroral displays and intense geomagnetic disturbances result. In particular, when a very gusty solar wind, generated by an intense solar flare, hits the magnetosphere, an intense geomagnetic storm occurs, considerably distorting the earth's magnetic field in the magnetosphere. As a result, the auroral oval expands unusually toward lower latitudes from its normal location in the polar region. For example, during the great geomagnetic storm of February 11, 1958, the oval shifted at times even south of the United States-Canadian border, so that the aurora temporarily disappeared from the polar sky. The Alaskan sky was almost completely deserted at about 1000 GMT on February 11, 1958. However, that situation did not last long. About half an hour later, auroras spread over a very

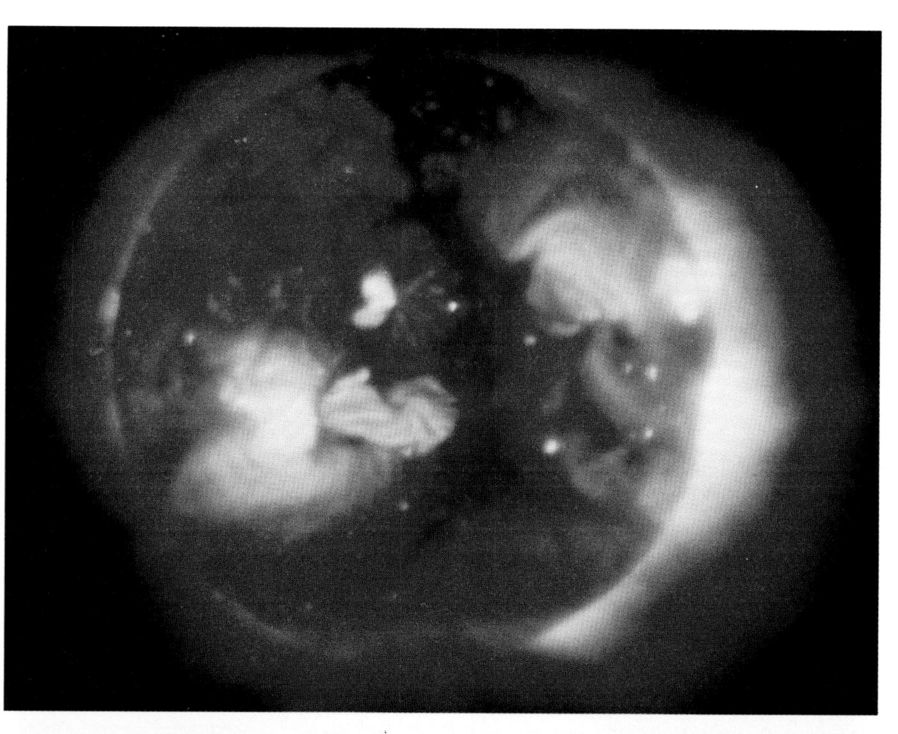

A soft x-ray photograph of the sun taken from Skylab. Solar physicists have discovered that a dark region in this photograph, called the "coronal hole," is a source of a continuous stream of high-speed solar wind. (NRL and NASA)

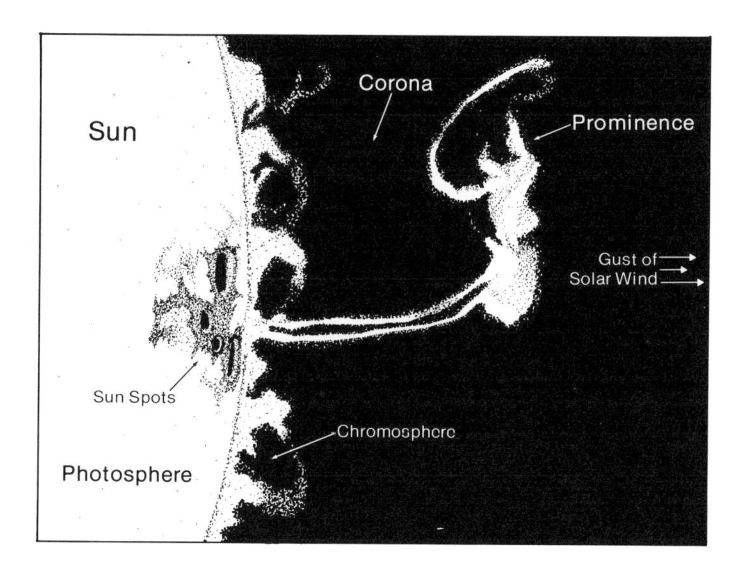

SOLAR FLARE: *Most intense solar flares tend to take place in a complex group of sunspots. A large prominence rises from such an active region, and a gusty solar wind is generated.*

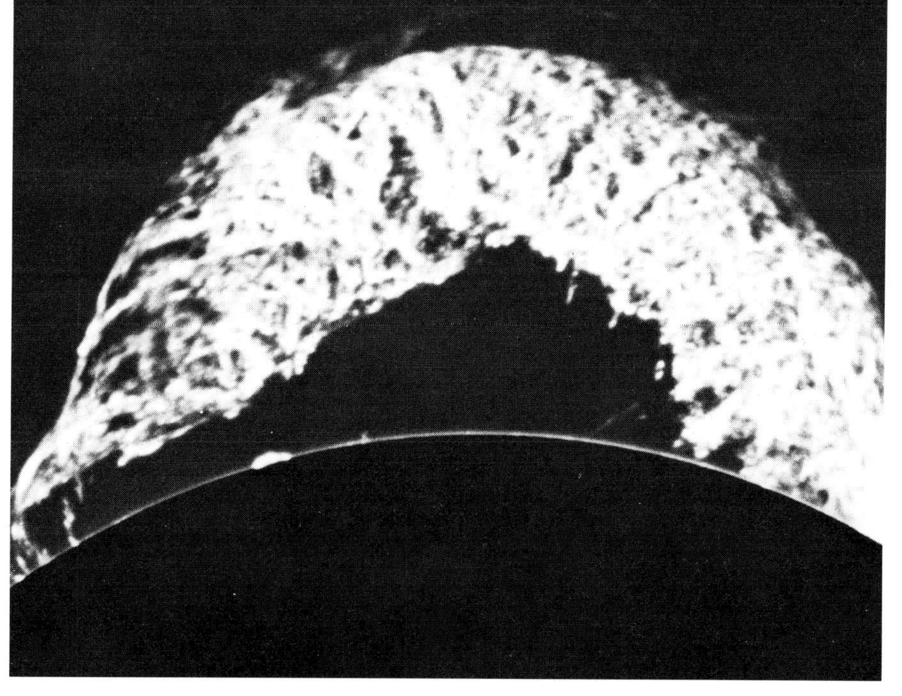

One of the largest prominences associated with an intense solar activity. (W.O. Roberts, High Altitude Observatory, NCAR)

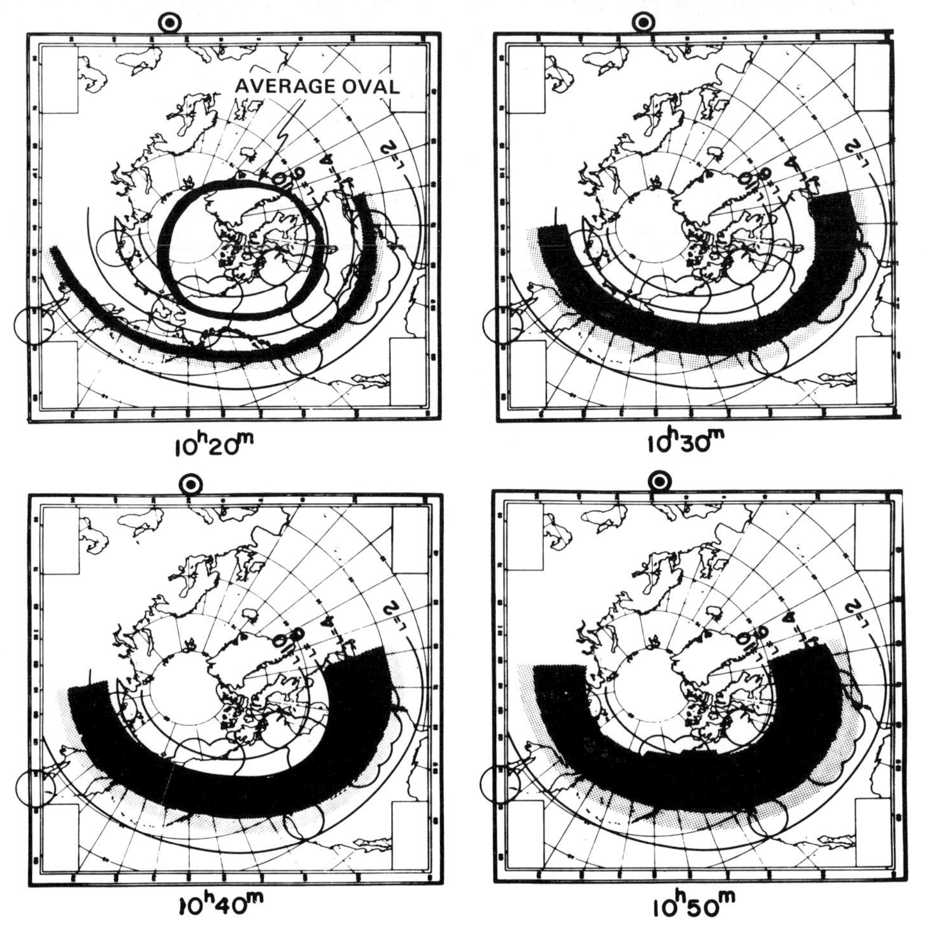

During an intense geomagnetic storm, the auroral oval expands toward lower latitudes from its normal location. During the great geomagnetic storm of February 11, 1958, the oval shifted at times even below the U.S.-Canadian border. These four views show the distribution of auroras at different epochs during the storm. The average size of the oval is indicated also in the first figure. (S.-I. Akasofu and S. Chapman, Journal of Atmospheric and Terrestrial Physics, *1962)*

The incoherent scatter radar, located at Chatanika, near Fairbanks, and operated by the Stanford Research Institute International. This radar is capable of measuring the electric field, electron density, ion and electron temperatures in the auroral ionosphere and is one of the most important tools in studying the aurora. (Stanford Research Institute International)

wide belt covering much of the northern United States and Canada.

The top of auroral curtains during such a great display tends to be as high as 1,000 kilometers (621 miles) and is a rich, dark red color, emitted by atomic oxygen. Because of its high altitude, the light can be seen from a very great distance, therefore from much farther south than usual. On February 11, 1958, the red aurora was seen from Mexico. Similar great auroral displays were seen on September 1, 1859, in Honolulu; February 4, 1872, from Bombay; September 25, 1909, in Singapore; May 13, 1921, in Samoa; and in Mexico on September 13 and 23, 1957. Again, it was this type of auroral display which provoked so much fear in medieval days.

For auroral scientists, auroral displays can be considered as an image in an oscilloscope which is a kind of TV tube or cathode-ray tube. An oscilloscope is used to diagnose electronic devices. In much the same way, scientists can "diagnose" the magnetosphere by studying the auroral "image." That is to say, they can infer characteristics of electric and magnetic disturbances which occur in space surrounding the earth from what they see displayed on the oscilloscope.

Auroral scientists also use a very large powerful tool to study the electrical storms in the magnetosphere and the ionosphere, a device called an incoherent scatter radar. The one located at Chatanika, Alaska, near Fairbanks, is shown above.

Why Study the Aurora?

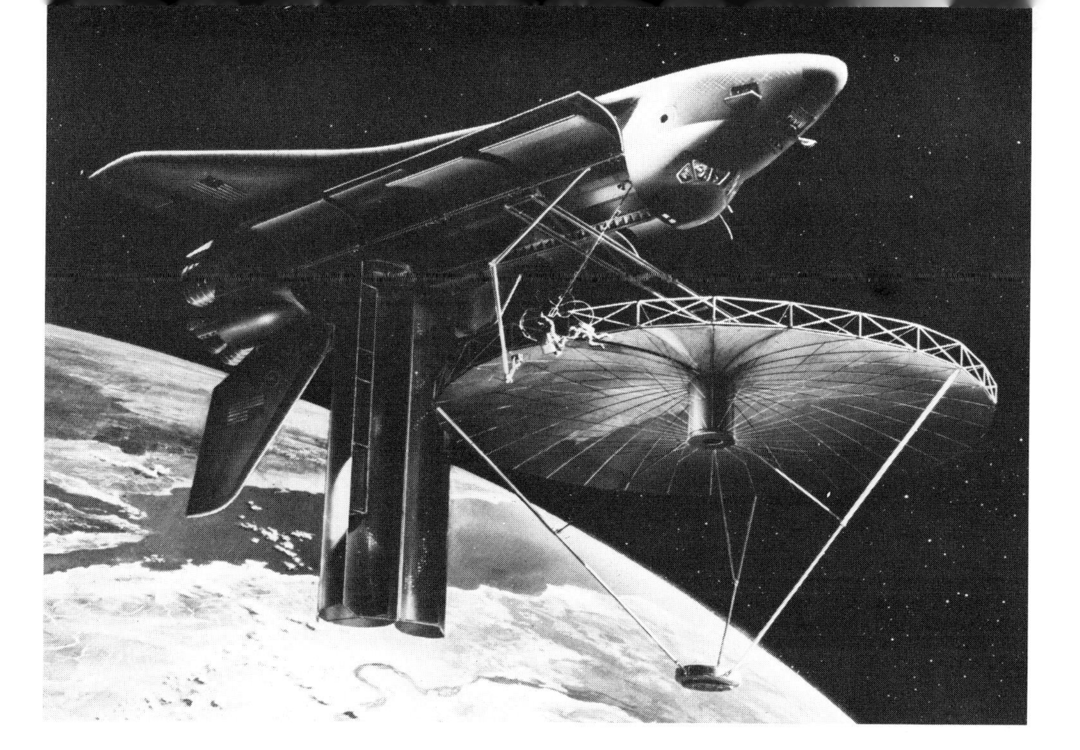

An artist's conception of the solar power station assembled in the sky by space shuttles. (NASA)

The first and easiest answer to the question posed is that most people want to know how such a magnificent phenomenon as the auroral lights is produced. As we have seen, auroral scientists have begun to unravel a chain of processes which cause the aurora. We know now at least that the aurora results from a large-scale electrical discharge surrounding the earth, powered by the solar wind-magnetosphere generator. Scientifically, however, our study of the aurora does not stop there. Auroral studies deal with extremely hot gases, so hot that the gases cannot remain neutral and thus consist of positively and negatively charged particles, called a plasma. Hannes Alfven noted that more than 99.9% of the matter in the universe is in the plasma state. Therefore, this extreme condition of matter is of fundamental importance to an understanding of almost every aspect of astronomy, such as the formation of the solar system, various processes in the sun, stars, quasi-stellar objects, pulsars and galaxies. The auroral plasma is the only one in nature that can be studied *directly* by satellite-borne instruments (except for information gathered by occasional visits of space probes to some of the other planets).

The term "plasma" for these extremely hot gases was coined by a famous American physicist, Irving Langmuir, who found that a group of charged particles exhibit a collective or organized behavior, called the plasma oscillation. It is, indeed, an aspect of that collective behavior of plasma which plays the role of the anode-cathode system (called the V-potential), accelerating auroral electrons. Without such an acceleration, we would not observe the aurora. Until just a few years ago, it had been firmly believed, as one of the basic principles among auroral physicists and astrophysicists, that auroral electrons could not be accelerated along magnetic field lines in a rarefied hot plasma. That this is not the case, as demonstrated by the presence of the aurora, will have far-reaching and revolutionary consequences in plasma physics and astrophysics, since many theories built upon that assumption must be revised completely. It is satisfying for auroral scientists that a study of the aurora has made such a fundamental contribution to science.

During the last 20 years, physicists have found evidence of considerably more unexpected behavior of plasma. Actually, it is that unexpected, unstable behavior of

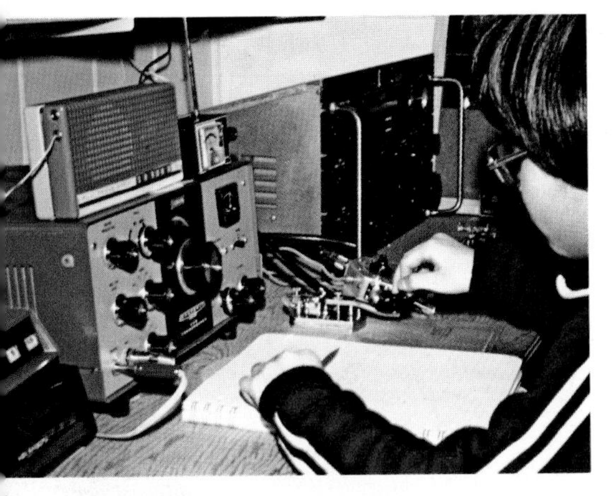

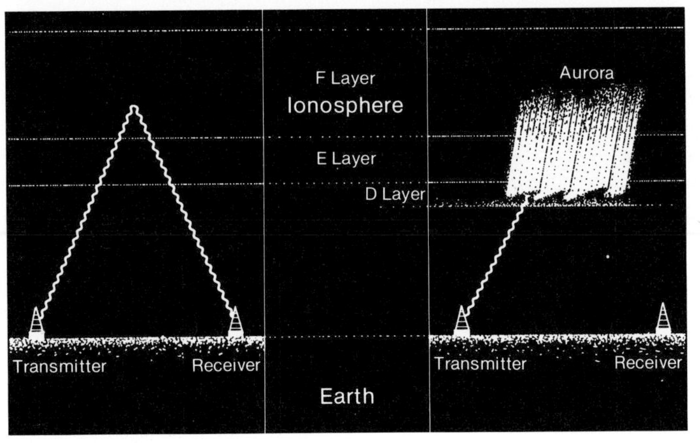

AURORAL EFFECTS ON THE IONOSPHERE AND RADIO WAVES TRAVERSING IT: *An abnormal layer of the ionosphere, called the D layer, is often formed during intense auroral displays, absorbing radio waves.*

Top — *Ham radio operators in high latitudes often cannot communicate with the rest of the world during auroral activity. (S.-I. Akasofu)*
Above — *Short wave radio communication is often disrupted during auroral displays as the aurora disturbs the ionosphere. (R. Hunsucker, Geophysical Institute, University of Alaska)*

plasma which at present is preventing us from achieving controlled thermonuclear fusion—most likely one of the main sources of our future energy.

Space electric power stations are now planned that will be placed at a distance of 38,000 kilometers (23,600 miles) from the earth, called the geosynchronous distance. However, these stations will be immersed in intense auroral electron beams and hot plasmas in the magnetosphere, and sometimes even in the solar wind during intense geomagnetic storms when the solar wind blows strong enough to squash the magnetosphere more than usual. The performance of space power stations in such a situation requires a great deal of study before they can safely be put into operation.

The aurora disturbs the polar ionosphere and the propagation of radio waves, causing disruption of radio communication and navigational difficulties. In particular, when an active auroral display is in progress, auroral electrons penetrate to a little below the E layer of the ionosphere, producing a particular type of ionosphere called the D layer. Radio waves (short waves) are absorbed there and thus will not be reflected back to reach a distant receiver. Ham radio operators in high latitudes often suffer from this effect of the aurora.

Auroral curtains in the E and F layers of the ionosphere tend to reflect radio waves, so that radar operation in high latitudes is often seriously disturbed. Satellite performance can also be affected by auroral electron beams and by high energy particles in the Van Allen radiation belts, which damage solar cells and cause damage to surface coatings. False commands can be produced by discharges as a result of abnormal electrostatic charging in a hot, tenuous, magnetospheric plasma; a satellite was even lost by such discharge effects.

Since the aurora is a discharge phenomenon, an electric current of about one million amperes flows along an auroral curtain in the ionospheric level, at about 100-110 kilometers (62-68 miles) in altitude. This ionospheric current is the major cause of geomagnetic disturbances in high latitudes, although changes in the direction of a compass needle are generally not more than a few degrees. Only very occasionally (during major geomagnetic storms) a compass needle will swing about 10 degrees for several minutes. When the intensity of this current varies, the associated magnetic field changes induce electric currents in long conductors on the earth, such as powerline systems, telegraph wires and oil or gas pipelines, causing transformer malfunction, unscheduled power outages and other damage. For example, the aurora seen by Carrington in 1859 caused considerable trouble in the telegraph system in Europe on that day. In 1897, Alfred Angot reported:

> At all the telegraphic stations in France the service was impeded during the whole of September 2. . . . The phenomenon consisted in a current producing continuous attraction of the armatures of the electro-magnets; . . . these currents were so strong that when the wire was isolated, and a conducting substance presented to it, it gave off vivid sparks. . . . The same day telluric currents were also observed in the greater part of the two hemispheres, in Switzerland, in Germany, in the British Isles, in North America, and throughout Australia. In the United States, in particular, they were so strong that for about two hours it was possible to send messages from Boston to Portland, and vice versa without any battery, using only the telluric current.

Alfred Angot, *The Aurora Borealis*,
D. Appleton, New York: 1897

Auroral curtain with a large-scale fold over Tromsø, Norway. (T. Berkey, courtesy of S.-I. Akasofu)

The aurora induces an intense electric current along lengthy conductors, such as oil pipelines, power transmission lines and telephone cables. This photograph shows the aurora and the trans-Alaska oil pipeline near Fairbanks. (S.-I. Akasofu)

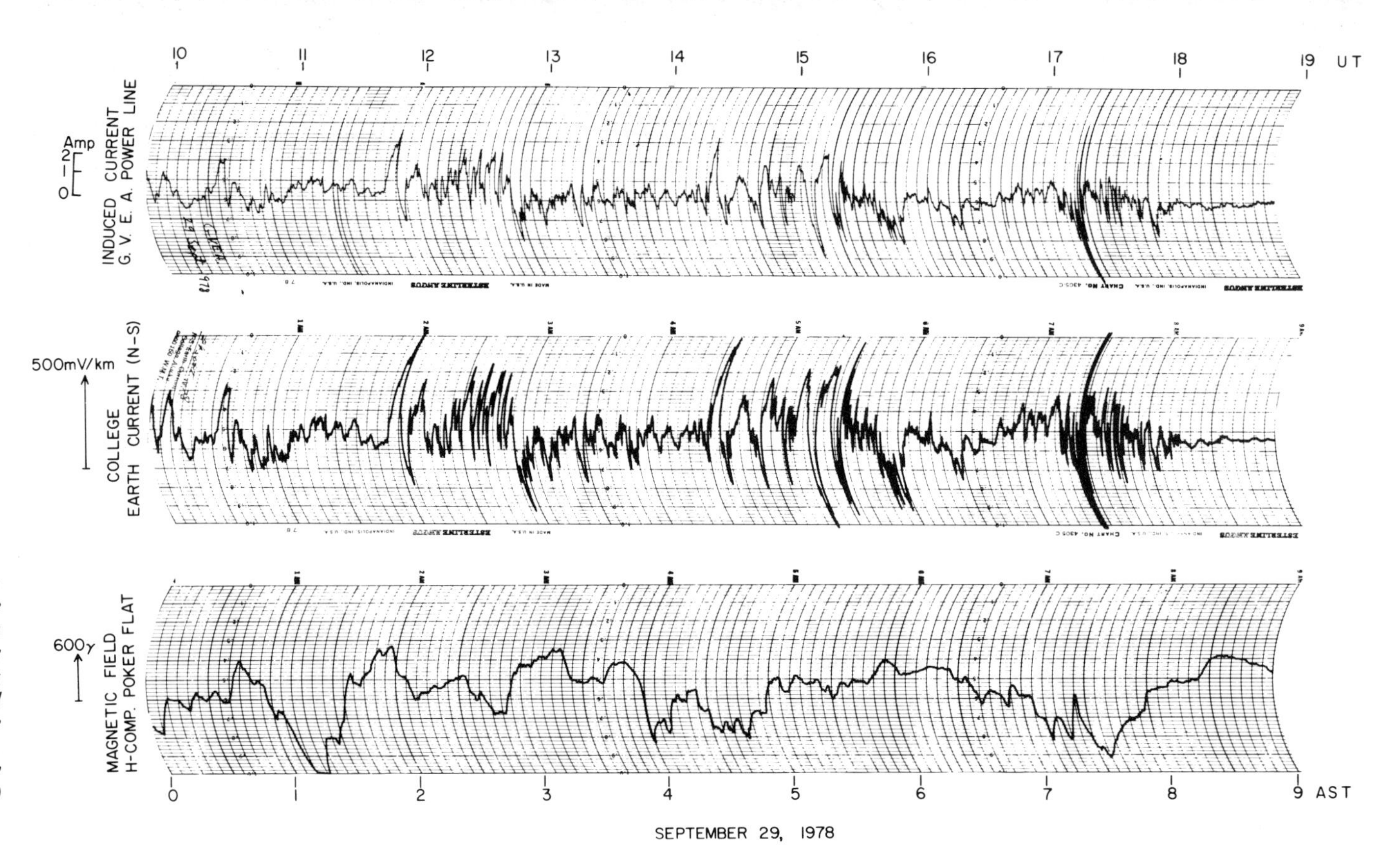

The power fluctuations recorded in the GVEA power transmission line from Healy to Fairbanks, Alaska (top), and the simultaneous auroral activity monitoring records, the electric field induced by the aurora (middle) and the magnetic disturbance (bottom). (Geophysical Institute, University of Alaska)

During the great red aurora of February 10, 1958, a temporary blackout occurred in northeastern Canada because of the tripping of circuit breakers in a transformer station. One of the more recent examples was the outage of a coaxial cable communication system in the Midwest during a violent geomagnetic disturbance in early August 1972. During the same period, Manitoba Hydro in Canada reported that the power flow being supplied by them to Minnesota dropped from 164 megawatts to 44 megawatts for about a minute, and later from 105 megawatts to an average value of 60 megawatts for 10 to 15 minutes. Also on the same day, Hydro Quebec reported a major power drop caused by an overload. In Newfoundland and Saskatchewan a considerable irregularity in the power was experienced and a transformer at British Columbia Hydro failed.

Auroral scientists are now working with solar physicists to provide reliable forecasts to power companies and communication facilities, as well as to the defense systems. There is a worldwide solar monitoring system that watches the sun continuously with various optical, radio and satellite-borne instruments. The actual forecast is issued from the Solar Forecasting Center, Space Environmental Laboratory, National Oceanographic and Atmospheric Administration, in Boulder, Colorado.

Because of these wide-ranging effects of the phenomenon, there are several research institutes in the world which dedicate their major efforts to studying the aurora. The Geophysical Institute, University of Alaska, is one of them. There are also a large number of research institutes and universities where many scientists are working on auroral phenomena.

ALASKA MERIDIAN CHAIN OF AURORAL OBSERVATORIES: *During the International Magnetosphere Study (an international cooperative study of the magnetosphere), the Geophysical Institute operated a magnetic meridian chain of observatories.*

A rocket launched from the Poker Flat Research Range, operated by the Geophysical Institute, University of Alaska. (Bill Baranauskas, Geophysical Institute, University of Alaska)

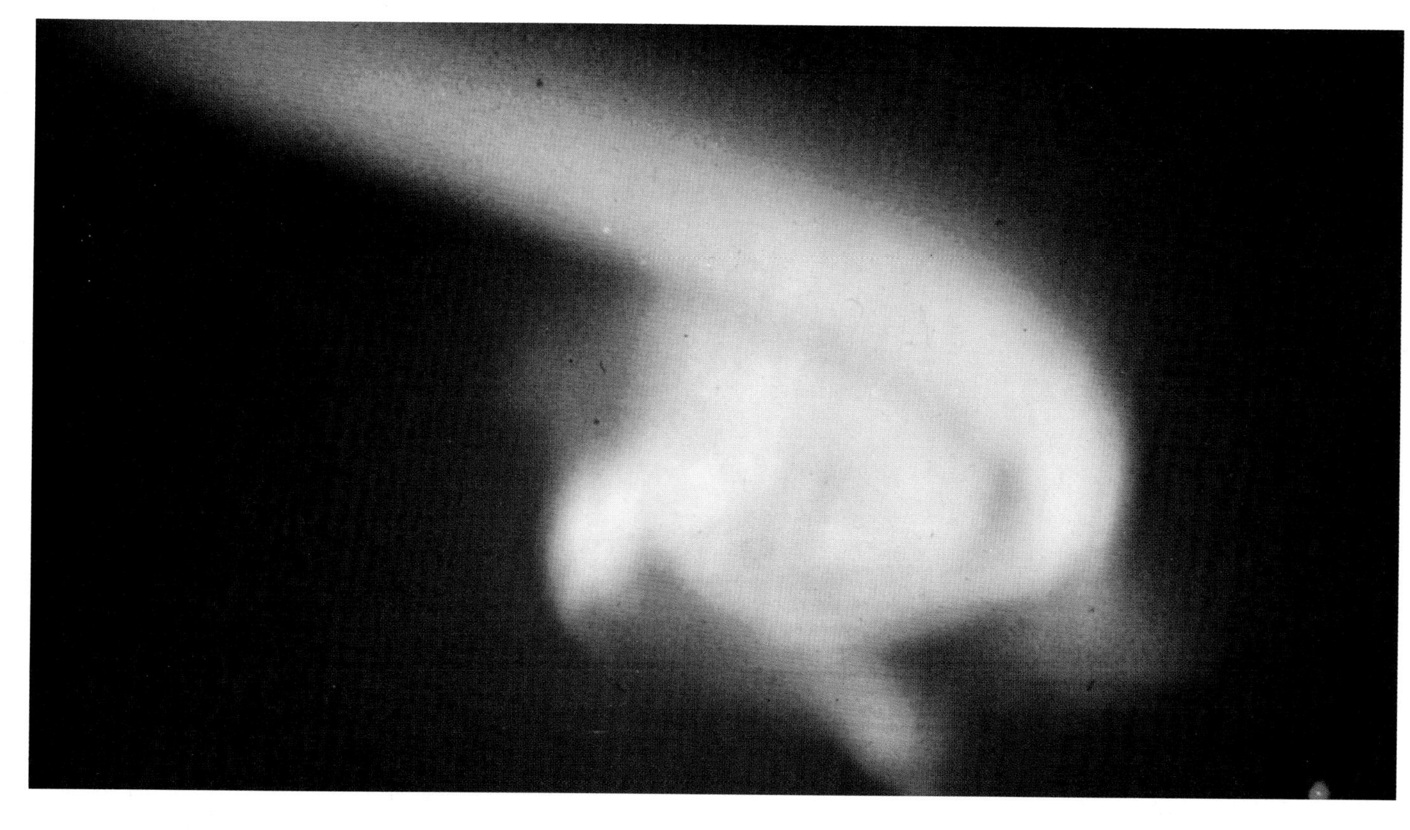

An active auroral curtain is developing in drape form. The crimson red color is also rich. (F. Yasuhara)

Auroral scientists believe that the aurora is a universal phenomenon which will take place wherever a high-speed plasma flow (such as the solar wind and stellar wind) interacts with a magnetized celestial body; many stars, including the sun, are magnetized. Some of the planets—Mercury and Jupiter, for example—are also magnetized, and in fact scientists have already found some indication of auroral phenomena (not visible features) in these planets. One of these days, a TV system aboard a Jovian probe will find the aurora there. Although Jupiter is known to be quite a different kind of planet from the earth, it will be satisfying to discover that both planets have something in common. Actually, it is already known that Jupiter has a gigantic magnetosphere. Some auroral scientists believe that basic plasma processes associated with solar flares are similar to those of auroral phenomena. Some also believe that by studying the auroral plasma we can better understand how the solar system evolved, since it must have been formed by contraction of a huge plasma cloud.

Harnessing aurora for power is still in the dream stage; we have not progressed very much from Thomas Knox's day in this respect. Some auroral scientists believe that auroral activity affects the weather and climate. This is one of the areas in which they have begun to work extensively.

Photographing the Aurora

To capture the northern lights on film, you will need the following camera equipment:

- ☐ A sturdy tripod.
- ☐ A locking-type cable release (some 35mm cameras have both *time* and *bulb* settings, but most have *bulb only*, which calls for use of the locking-type cable release).
- ☐ A camera with an f/3.5 lens (or faster).

It is best to photograph the lights on a night when they are not moving too rapidly. And, as a general rule, photos improve if you manage to include recognizable subjects in the foreground—trees and lighted cabins being favorites of many photographers. Set your camera up at least 75 feet back from the foreground objects to make sure that both the foreground and aurora are in sharp focus.

Normal and wide-angle lenses are best. Try to keep your exposures under a minute — 10 to 30 seconds are generally best. The following lens openings and exposure times are only a starting point, since the amount of light generated by the aurora is inconsistent. (It's best to bracket exposures widely for best results.)

	ASA 200	ASA 400
f1.2	3 sec.	2 sec.
f1.4	5	3
f1.8	7	4
f2	20	10
f2.8	40	20
f3.5	60	30

Ektachrome 200 and 400 color films can be push-processed in the home darkroom or by some custom-color labs, allowing use of higher ASA ratings (800, 1200 or even 1600 on the 400 ASA film, for example). Kodak will push-process film if you include an ESP-1 envelope with your standard film-processing mailer. (Consult your local camera store for details.)

A few notes of caution:

- ☐ Remember to protect the camera from low temperatures until you are ready to make your exposures. Some newer cameras, in particular, have electrically controlled shutters that will not function properly at low temperatures.
- ☐ Wind the film slowly to reduce the possibility of static electricity, which can lead to streaks on the film. Grounding the camera when rewinding can help prevent the static-electricity problem. (To ground the camera, hold it against a water pipe, drain pipe, metal fence post or other grounded object.)

Follow the basic rules, experiment with exposures, and you should obtain good results.

When the temperature is below -20° F, you must use extra caution when operating your camera. This photo shows that the shutter was not working properly. (S.-I. Akasofu)

Additional Reading

Akasofu, Syun Ichi, *Polar and Magnetospheric Substorms,* Dordrecht, Holland: D. Reidel Publ. Co., 1968.

Akasofu, Syun-Ichi and Chapman, Sydney, *Solar-Terrestrial Physics,* London: Oxford University Press, 1972.

Alfven, Hannes, *Cosmical Electrodynamics*, London: Oxford University Press, 1950.

Beecher, Arthur, "When the Aurora Hits the Earth," *The ALASKA SPORTSMAN®*, November, 1939.

Chamberlain, Joseph W., *Physics of the Aurora and Airglow*, New York: Academic Press, 1961.

Cresswell, G., "Fire in the Sky," *ALASKA SPORTSMAN®*, January, 1968.

Damron, Floyd, "How to Photograph the Northern Lights," *ALASKA®* magazine, November, 1973.

Ellison, M. A., *The Sun and Its Influence,* Third edition, London: Routledge and Kegan Paul Ltd., 1968.

Elvey, C. T., "Can You Hear the Northern Lights?" *ALASKA SPORTSMAN®*, June, 1962.

Hunsucker, Robert D., "The Northern Lights," *ALASKA SPORTSMAN®*, March, 1963.

Jones, Alister Vallance, *Aurora*, D. Reidel Publ. Co., Dordrecht, Holland, 1974.

Petric, William, *Keoeeit - The Story of the Aurora Borealis,* Pergamon Press, Oxford, 1963.

Stormer, Carl, *The Polar Aurora*, London: Oxford University Press, 1955.

Zirin, Harold, *The Solar Atmosphere*, Waltham, Mass.: Blaisdell Publ. Co, 1966.

Alaska Geographic® Back Issues

The North Slope, Vol. 1, No. 1. The charter issue of *ALASKA GEOGRAPHIC®.* Out of print.

One Man's Wilderness, Vol. 1, No. 2. The story of a dream shared by many, fulfilled by a few; a man goes into the Bush, builds a cabin and shares his incredible wilderness experience. Color photos. 116 pages, $9.95.

Admiralty . . . Island in Contention, Vol. 1, No. 3. An intimate and multifaceted view of Admiralty; it's geological and historical past, its present-day geography, wildlife and sparse human population. Color photos. 78 pages, $5.

Fisheries of the North Pacific: History, Species, Gear & Processes, Vol. 1, No. 4. Out of print. (Book edition available)

The Alaska-Yukon Wild Flowers Guide, Vol. 2, No. 1. Out of print. (Book edition available)

Richard Harrington's Yukon, Vol. 2, No. 2. Out of print.

Prince William Sound, Vol. 2, No. 3. Out of print.

Yakutat: The Turbulent Crescent, Vol. 2, No. 4. Out of print.

Glacier Bay: Old Ice, New Land, Vol. 3, No. 1. The expansive wilderness of southeastern Alaska's Glacier Bay National Monument (recently proclaimed a national park and preserve) unfolds in crisp text and color photographs. Records the flora and fauna of the area, its natural history, with hike and cruise information, plus a large-scale color map. 132 pages, $11.95.

The Land: Eye of the Storm, Vol. 3, No. 2. Out of print.

Richard Harrington's Antarctic, Vol. 3, No. 3. The Canadian photojournalist guides readers through remote and little understood regions of the Antarctic and subantarctic. More than 200 color photos and a large fold-out map. 104 pages, $8.95.

The Silver Years of the Alaska Canned Salmon Industry: An Album of Historical Photos, Vol. 3, No. 4. Out of print.

Alaska's Volcanoes: Northern Link in the Ring of Fire, Vol. 4, No. 1. Out of print.

The Brooks Range: Environmental Watershed, Vol. 4, No. 2. Out of print.

Kodiak: Island of Change, Vol. 4, No. 3. Out of print.

Wilderness Proposals: Which Way for Alaska's Lands? Vol. 4, No. 4. Out of print.

Cook Inlet Country, Vol. 5, No. 1. Out of print. All-new edition (Vol. 10, No. 2) available.

Southeast: Alaska's Panhandle, Vol. 5, No. 2. Explores southeastern Alaska's maze of fjords and islands, mossy forests and glacier-draped mountains — from Dixon Entrance to Icy Bay, including all of the state's fabled Inside Passage. Along the way are profiles of every town, together with a look at the region's history, economy, people, attractions and future. Includes large fold-out map and seven area maps. 192 pages, $12.95.

Bristol Bay Basin, Vol. 5, No. 3. Out of print.

Alaska Whales and Whaling, Vol. 5, No. 4. The wonders of whales in Alaska — their life cycles, travels and travails — are examined, with an authoritative history of commercial and subsistence whaling in the North. Includes a fold-out poster of 14 major whale species in Alaska in perspective, color photos and illustrations, with historical photos and line drawings. 144 pages, $12.95.

Yukon-Kuskokwim Delta, Vol. 6, No. 1. Out of print.

The Aurora Borealis, Vol. 6, No. 2. The northern lights — in ancient times seen as a dreadful forecast of doom, in modern days an inspiration to countless poets. What causes the aurora, how it works, how and why scientists are studying it today and its implications for our future. 96 pages, $7.95.

Alaska's Native People, Vol. 6, No. 3. Examine the varied worlds of the Inupiat Eskimo, Yup'ik Eskimo, Athabascan, Aleut, Tlingit, Haida and Tsimshian. Included are sensitive, informative articles by Native writers, plus a large, four-color map detailing the Native villages and defining the language areas, 304 pages, $24.95.

The Stikine, Vol. 6, No. 4. River route to three Canadian gold strikes in the 1800s, the Stikine is the largest and most navigable of several rivers that flow from northwestern Canada through southeastern Alaska on their way to the sea. Illustrated with contemporary color photos and historic black-and-white; includes a large fold-out map. 96 pages, $9.95.

Alaska's Great Interior, Vol. 7, No. 1. Alaska's rich Interior country, west from the Alaska-Yukon Territory border and including the huge drainage between the Alaska Range and the Brooks Range, is covered thoroughly. Included are the region's people, communities, history, economy, wilderness areas and wildlife. Illustrated with contemporary color and black-and-white photos. Includes a large fold-out map. 128 pages, $9.95.

A Photographic Geography of Alaska, Vol. 7, No. 2. An overview of the entire state — a visual tour through the six regions of Alaska: Southeast, Southcentral/Gulf Coast, Alaska Peninsula and Aleutians, Bering Sea Coast, Arctic and Interior. Plus a handy appendix of valuable information — "Facts About Alaska." Revised in 1983. Approximately 160 color and black-and-white photos and 35 maps. 192 pages, $15.95.

The Aleutians, Vol. 7, No. 3. Home of the Aleut, a tremendous wildlife spectacle, a major World War II battleground and now the heart of a thriving new commercial fishing industry. Contemporary color and black-and-white photographs, and a large fold-out map. 224 pages, $14.95.

Klondike Lost: A Decade of Photographs by Kinsey & Kinsey, Vol. 7, No. 4. An album of rare photographs and all-new text about the lost Klondike boom town of Grand Forks, second in size only to Dawson during the gold rush. $12.95.

Wrangell-Saint Elias, Vol. 8, No. 1. Mountains, including the continent's second- and fourth-highest peaks, dominate this international wilderness that sweeps from the Wrangell Mountains in Alaska to the southern Saint Elias range in Canada. Includes a large fold-out map. 144 pages, $9.95.

Alaska Mammals, Vol. 8, No. 2. From tiny ground squirrels to the powerful polar bear, and from the tundra to the magnificent whales inhabiting Alaska's waters, this volume includes 80 species of mammals found in Alaska. 184 pages, $12.95.

The Kotzebue Basin, Vol. 8, No. 3. Examines northwestern Alaska's thriving trading area of Kotzebue Sound and the Kobuk and Noatak river basins, lifelines of the region's Inupiat Eskimos, early explorers, and present-day, hardy residents. 184 pages, $12.95.

Alaska National Interest Lands, Vol. 8, No. 4. Following passage of the bill formalizing Alaska's national interest land selections (d-2 lands), longtime Alaskans Celia Hunter and Ginny Wood review each selection, outlining location, size, access, and briefly describing the region's special attractions. 242 pages, $14.95.

Alaska's Glaciers, Vol. 9, No. 1. Examines in depth the massive rivers of ice, their composition, exploration, present-day distribution and scientific significance. 144 pages, $9.95.

Sitka and Its Ocean/Island World, Vol. 9, No. 2. From the elegant capital of Russian America to a beautiful but modern port, Sitka, on Baranof Island, has become a commercial and cultural center for southeastern Alaska. 128 pages, $9.95.

Islands of the Seals: The Pribilofs, Vol. 9, No. 3. Great herds of northern fur seals drew Russians and Aleuts to these remote Bering Sea islands where they founded permanent communities and established a unique international commerce. 128 pages, $9.95.

Alaska's Oil/Gas & Minerals Industry, Vol. 9, No. 4. Experts detail the geological processes and resulting mineral and fossil fuel resources that are now in the forefront of Alaska's economy. Illustrated with historical black-and-white and contemporary color photographs. 216 pages, $12.95.

Adventure Roads North: The Story of the Alaska Highway and Other Roads in *The MILEPOST®*, Vol. 10, No. 1. From Alaska's first highway — the Richardson — to the famous Alaska Highway, first overland route to the 49th state, text and photos provide a history of Alaska's roads and take a mile-by-mile look at the country they cross. 224 pages, $14.95.

ANCHORAGE and the Cook Inlet Basin, Vol. 10, No. 2. "Anchorage country" . . . the Kenai, the Susitna Valley, and Matanuska. Heavily illustrated in color and including three illustrated maps . . . one an uproarious artist's forecast of "Anchorage 2035." 168 pages, $14.95.

Alaska's Salmon Fisheries, Vol. 10, No. 3. The work of *ALASKA®* magazine Outdoors Editor Jim Rearden, this issue takes a comprehensive look at Alaska's most valuable commercial fishery. 128 pages, $12.95.

Up the Koyukuk, Vol. 10, No. 4. Highlights the Koyukuk region of north-central Alaska . . . the wildlife, fauna, Native culture and more. 152 pages. $14.95.

Nome: City of the Golden Beaches, Vol. 11, No. 1. The colorful history of Alaska's most famous gold rush town has never been told like this before. Illustrated with hundreds of rare black-and-white photos, the book traces the story of Nome from the crazy days of the 1900 gold rush. 184 pages, $14.95.

Alaska's Farms and Gardens, Vol. 11, No. 2. An overview of the past, present, and future of agriculture in Alaska, and a wealth of information on how to grow your own fruit and vegetables in the north. 144 pages, $12.95.

Chilkat River Valley, Vol. 11, No. 3. This issue explores the mountain-rimmed valley at the head of the Inside Passage, its natural resources, and those hardy residents who make their home along the Chilkat. 112 pages, $12.95.

Alaska Steam, Vol. 11, No. 4. A pictorial history of the Alaska Steamship Company pioneering the northern travel lanes. Compiled by Lucile McDonald. More than 100 black-and-white historical photos. 160 pages. $12.95.

Northwest Territories, Vol. 12, No. 1. An in-depth look at some of the most beautiful and isolated land in North America. Compiled by Richard Harrington. 148 color photos. 136 pages. $12.95.

Your $30 membership in the Alaska Geographic Society includes 4 subsequent issues of *ALASKA GEOGRAPHIC®*, the Society's official quarterly. Please add $4 for non-U.S. membership.

Additional membership information available upon request. Single copies of the *ALASKA GEOGRAPHIC®* back issues are also available. When ordering, please make payments in U.S. funds and add $1 postage/handling per copy. To order back issues send your check or money order and volumes desired to:

The Alaska Geographic Society

Box 4-EEE, Anchorage, Alaska 99509